Victor Farias Moebus
Luiz Antonio Moura Keller

Fermenting Microorganisms in the Removal of Contaminates in Milk

Victor Farias Moebus
Luiz Antonio Moura Keller

Fermenting Microorganisms in the Removal of Contaminates in Milk

Evaluation of Adsorption Capacity and Biometabolization

ScienciaScripts

Cover image: www.ingimage.com

This book is a translation from the original published under ISBN 978-620-6-75707-8.

Publisher:
Sciencia Scripts
is a trademark of
Dodo Books Indian Ocean Ltd. and OmniScriptum S.R.L publishing group

120 High Road, East Finchley, London, N2 9ED, United Kingdom
Str. Armeneasca 28/1, office 1, Chisinau MD-2012, Republic of Moldova, Europe
Printed at: see last page
ISBN: 978-620-3-61660-6

ACKNOWLEDGEMENTS

To the Fluminense Federal University, the Laboratory of Physical-Chemical Control of Products of Animal Origin, the Laboratory of Microbiological Control of Products of Animal Origin, the Department of Food Technology - MTA of the Veterinary Faculty of the UFF.
To the Postgraduate Programme in Veterinary Medicine at UFF, for their support during the research.
To the Rio de Janeiro State Research Foundation (FAPERJ) for the grant awarded for my studies.
To my advisor, Luiz Antonio Moura Keller, for being more than an advisor and for always being present and willing to help us in any situation.
To Professor Eliane Teixeira Mársico and the other professors at the Laboratory for the Physical and Chemical Control of Products of Animal Origin, for welcoming us with great affection and integrating us into the team, making it possible to carry out this work.
To Professors Robson Maia Franco and Maria Carmela Kasnowski Holanda Duarte of the Laboratory for the Microbiological Control of Animal Products, for the knowledge they passed on, for their advice on microbiology and for always being available when I needed help.
To Professor Kelly Moura Keller of the Federal University of Minas Gerais, for all the help she received and for her willingness to come to Rio de Janeiro to help me at a crucial moment in my work.
To Professors Vanessa Naciuk Castelo Branco, Alice Goncalves Martins Gonzalez and Leandro Machado Rocha from the Faculty of Pharmacy of the Fluminense Federal University, for all the support and reagents lent to carry out this work.
To Professor Eliana Mesquita, for her support, encouragement, tugs on the ear; for the conversations, sweets and jokes.

To my great friend Pedro Borghi, for helping me acquire the materials I needed for the work.
To my mother Fátima Moebus and my grandmother Alcinéa Moebus, for always waiting for me with a lot of love and affection and making me forget my problems when I'm in Petrópolis, I love you.
To my dear Camilla Queiroz, for all your love, companionship and support. Thank you for being the best person in the world, for your patience and affection in the most crucial moments of this period. Thank you for loving me, for always being by my side, supporting my crazy behaviour and for choosing to start a family with me. I love you more than anything.
To the best dog in the world, Laura sausage, for showing me how good it is to take care of a life and making me work hard to buy you a toy. As well as keeping up with my CrossFit running after toys.
To my brother Matheus Moebus and my sister-in-law Isabela Telles, for their support, company, outbursts during games of ludo and marvellous recipes.
A very special thank you to my friends Leonardo Pinto and Felipe Köptcke for all their support. Without the sleepless nights, revising and helping with articles, arriving early and leaving the lab late and your help, this work would not have happened in this way. Thank you very much.

To my Project friends, Beatriz Clarissa and Tatiana Reis, for all their in-person and long-distance support during the project.

To our friends Marcos Aronovich and Christianne Perali, for adopting us and being a family away from home.

To our dearest friends Nathalia (Arthur's mum) and Arthur Keller, for always welcoming us and bringing us moments of relaxation and fun.

To my undergraduate friends, for being with me every step of the way throughout all these years.

To my friends in Petrópolis who, despite the distance and absence, continue to be with me at all times.

To the entire team at the Laboratory for the Physical and Chemical Control of Products of Animal Origin, for all the support, encouragement, laughter and silly things said during moments of work and relaxation.

To my colleagues at the Fish Laboratory and the Toxicology and Animal Health Centre, for all their companionship.

To my fellow postgraduates, for their conviviality and companionship during the dissertations.

"Consider your origin. You were not formed to live like brutes, but to follow virtue and knowledge"

Dante Alighieri

SUMMARY

Milk and its derivatives are one of the main components of the national livestock industry and have a significant impact on the economy. However, the production chain for these products is susceptible to a number of flaws, especially with regard to contamination by undesirable substances. These include residues of antimicrobial drugs and mycotoxins, which can trigger pathogenic conditions in humans and lead to technological complications during the production of by-products. Due to the thermal resistance of mycotoxins, various technologies have been developed over the years to minimise or eliminate these compounds in both livestock and milk. Among these technologies, the use of fermenting microorganisms such as Saccharomyces cerevisiae and lactic acid bacteria (LAB) for the adsorption or biotransformation of mycotoxins in food matrices has been widely adopted. Probiotic microorganisms, specifically yeasts and lactic acid bacteria, play a crucial role in these technologies due to their recognised safety, remarkable adsorption capacity and beneficial impact on the host's health, allowing them to be incorporated into food formulations and diets. Aflatoxin B1 (AFB1) and zearalenone (ZEA) represent some of the most common and significant mycotoxins in agricultural products, with AFB1 being a hepatocarcinogen and ZEA responsible for reproductive dysfunctions. In the first study conducted, the aim was to assess the in vitro detoxification of AFB1 and ZEA at pH 3 and 6 by three strains of S. cerevisiae isolated from bovine fodder (LL74, LL08 and LL83), analysing their adsorption and biotransformation capacities. The strains evaluated showed high mycotoxin adsorption rates (20-55%), especially the LL08 strain, although they showed low biotransformation, resulting in similar performance to the commercial strain. The results suggest that pH does not affect the detoxification of AFB1, while the use of S. cerevisiae strain LL83 at 20 mg/mL led to the highest adsorption of ZEA at pH 3. These strains indicate potential for in vivo application and high attractiveness for commercial applications, with strain LL08 standing out as the most promising. The second study aimed to investigate and characterise the potential of commercial human probiotic strains for biotechnological applications in milk. Twenty strains were evaluated in terms of growth kinetics, gastrointestinal assays, lactose consumption, antimicrobial resistance and the ability to bioremove aflatoxin M1 (AFM1) and antimicrobial residues. After initial screening, five strains of Lactobacillus spp. were selected as potential biotechnological agents, showing adequate concentrations in 24 hours, resistance to antimicrobials, good intestinal adhesion and high lactose consumption. Although biometabolisation of AFM1 and antimicrobials did not occur, the strains showed high adsorption rates of AFM1, reducing its bioavailability in food matrices. The selected strains have fundamental properties for biotechnological applications in milk and potential for decontamination, allowing them to be applied in bioremediation technologies, especially in the fermentation of low-lactose products. This study opens up perspectives for the discovery and utilisation of new strains with potential applicability in the removal of contaminants from food matrices, in the food industry and in the production of secondary metabolites of industrial interest.

Keywords: Lactic acid bacteria, Saccharomyces cerevisiae, antimicrobials, biotransformation

SUMMARY

1 INTRODUCTION

Milk is essential for human nutrition and is produced all over the world. It is an abundant and cheap source of animal protein on the Brazilian market and is easy to obtain. The production of milk and dairy products in Brazil is important on a social, economic and environmental level, generating millions of jobs throughout the production chain, with an economic return in the order of millions per year (ALMEIDA NETO et al., 2017).

However, milk and its derivatives can be associated with various types of contamination, whether of microbiological or chemical origin, which, due to the lack of monitoring and reporting of outbreaks, can cause serious harm to the population (GERMANO; GERMANO, 2019).

Among the substances that can cause harm to the consumer, two classes of substance deserve to be highlighted due to the seriousness of the effects they cause: mycotoxins and antimicrobial residues.

Mycotoxins are secondary metabolites produced by some filamentous fungi, which are observed in both plant and animal products and can be detected at various levels of the food chain, their production being favoured by environmental stress conditions such as humidity and temperature (DI CASTRO, 2015).

Antimicrobials are compounds that reduce or inhibit microbial growth when used in optimal inhibitory concentrations (CARNEIRO, 2011). In Veterinary Medicine, antimicrobials are widely used in all stages of production or the life cycle of animals, especially in mammary gland infections (mastitis) and respiratory tract diseases (RONQUILLO; HERNANDEZ, 2017).

Therefore, even when correctly applying the necessary control measures to minimise the incidence of chemical contaminants in milk, the use of removal technologies based on substances capable of removing chemical contaminants from food and from the animal's gastrointestinal tract after ingestion have been widely employed (ADEGBEYE et al., 2020; COLOVIĆ et al., 2019).

To remove mycotoxins, anti-mycotoxin additives, whether inorganic or organic, have been used in various animal species, with results varying according to the animal species, the additive used and the substance to be removed (DALLMANN et al., 2021; REIS et al., 2020).

Despite their high efficacy in removing aflatoxins, inorganic adsorbents do not have the same potential for other mycotoxins. This, combined with the low The specificity of the binding sites has increasingly motivated the use of microorganisms to mitigate mycotoxins. Increasingly, the removal of various mycotoxins in food has been based on their interactions with fermenting microorganisms, mainly through biodegradation and bioadsorption mechanisms (KELLER et al., 2015; LI et al., 2018). Among the microorganisms used to mitigate mycotoxins are yeasts, especially Saccharomyces cerevisiae, and bacteria, especially lactic acid bacteria due to their probiotic potential.

Yeasts, whether viable or not, have a high adsorption capacity and are able to reduce the bioavailability of various mycotoxins in food and feed by binding the mycotoxins to the components of their cell walls (PFLIEGLER; PUSZTAHELYI; PÓCSI, 2015). Precisely because they are bioadsorbents that have been widely described in the literature, prior knowledge of the methodologies involved in these bioprocesses is capable of subsidising knowledge when applied to other groups of bacteria with bioremoval potential. The group of lactic acid bacteria is made up of Gram-positive bacteria with varied morphology. Lactic acid bacteria are widespread in nature, even in unfavourable environments. They are used to produce fermented foods, meats, vegetables, drinks, fruit and as probiotics (FORSYTHE, 2002; OLIVEIRA, 2009).

As a complement to mycotoxin removal, biodegradation technology is based on the use of microorganisms and/or enzymes to degrade mycotoxins into inert or less toxic compounds, depending on the metabolic pathway of the living microorganisms. However, research into the biodegradation of mycotoxins by probiotics is very limited (LI et al., 2018; LIU; XIE; WEI, 2022; SARLAK et al., 2017).

Unlike the removal of mycotoxins from food matrices, the removal of antimicrobials by microorganisms has been little explored, setting a precedent for research in this area.

The aim of this study is to assess the mitigation potential of probiotic microorganisms against mycotoxins and antimicrobial residues, evaluating the adsorptive and biodegradation potential of microorganisms in vitro and in raw milk.

2 THEORETICAL BACKGROUND

2.1 MILK

Milk is essential for human nutrition and is produced all over the world. Its importance can be seen in the global production and economic environment, especially in developing countries and in family farming systems. Milk is an abundant and cheap source of animal protein on the Brazilian market and is easy to obtain. The production of milk and dairy products in Brazil is important on a social and economic level, as it generates millions of jobs throughout the production chain, with an economic return in the order of millions per year, surpassing other sectors (ALMEIDA NETO et al., 2017). In the state of Rio de Janeiro alone, due to the strong presence of the dairy chain there, more than 120 million litres of milk were produced in the 1st quarter of 2023 alone (IBGE, 2023).

Over the last three decades, world milk production has increased by more than 50%, reaching 906 million tonnes in 2020. In the same year, world milk production grew by around 2.0%, as did the production of its derivatives, which grew by 1.2% in recent years when considering the largest producers (FAO, 2021). The projection of this scenario for the next decade is that there will be a significant increase in milk production of around 1.7% per year. Consequently, there will be a significant increase in the production and development of dairy products (FAO, 2019). According to the Ministry of Health's recommendations, milk consumption, in fluid form or in the form of dairy products, varies according to age. The recommendation for children up to 10 years old is 400 mL/day, i.e. 146 litres/year of fluid milk or equivalent in the form of dairy products. For young people aged 11 to 19, consumption is higher, at 700 mL/day or 256 litres/year and for adults over 20 the recommendation is 600 mL/day or 219 litres/year, including for the elderly, but consumption for this group of people should be mainly skimmed (BRASIL, 2013).

Milk and its derivatives are a group of foods with great nutritional value, as they are considerable sources of proteins with high biological value, as well as vitamins and minerals (BRASIL, 2014). In Brazil, they can be found on the table of practically every household and this is reflected in consumption, where around 173 litres were consumed per capita between 2019 and 2021 (BRASIL, 2021).

However, milk and its derivatives can be associated with various types of contamination, whether of microbiological or chemical origin, which, due to the lack of monitoring and reporting of outbreaks, can cause serious harm to the population (GERMANO; GERMANO, 2019).

2.1.1 **Legislation**

According to the Regulations for the Industrial and Sanitary Inspection of Products of Animal Origin (RIISPOA), milk, without any other specification, is understood to be the product derived from the complete and uninterrupted milking, under hygienic conditions, of healthy,

well-fed and rested cows (BRASIL, 2017).
Immediately after milking, the raw milk is refrigerated on the property and sent to dairy processing and dairy product production establishments, where it is known as refrigerated raw milk (BRASIL, 2017).
Since raw milk is not subjected to any preservation process, the temperature at which it is transported and packaged is important for guaranteeing the quality of the product. It should not exceed 9°C on receipt and should be stored at up to 4°C in order to avoid changes in the milk's characteristics (BRASIL, 2018).
Normative Instruction 76 of 26 November 2018 also sets out the other requirements for characterising milk within the product's standard of identity and quality. These include sensory parameters, such as colour and odour, and physicochemical parameters, such as the minimum levels of lipids, protein, lactose and solids, as well as acidity, cryoscopy and alizarol stability (BRASIL, 2018).
Since raw milk should not be consumed without prior processing, Normative Instruction No. 60 of 23 December 2019 does not state the limits of microorganisms allowed in the product, which are only present in pasteurised products (BRASIL, 2019). However, raw milk must have a maximum limit of 9×10^5 CFU/mL before it is processed in the processing plant. As this is done through standardised plate counts using non-selective culture media, the survey only provides a quantitative value of microbial contamination (BRASIL, 2018).

2.2 CHEMICAL CONTAMINANTS IN MILK

According to the Technical Regulation on Identity and Quality (RTIQ), refrigerated raw milk must not contain any substances foreign to its composition, such as agents that inhibit microbial growth, neutralise acidity and reconstitute density or the cryoscopic index (BRASIL, 2018). With the exception of agents that inhibit microbial growth, the other contaminants mentioned are easily detected through the physicochemical control techniques used in routine evaluations (BRASIL, 2018; IAL, 2008).
The purpose of adding chemical substances to milk is to correct some parameter so that the product conforms to what is recommended, altering its composition. This practice, when used, is characterised as product fraud and can cause damage to consumer health. In addition to the economic losses generated by reduced yields and altered quality of the processed products (CORTEZ et al., 2010; PANCIERE; RIBEIRO, 2021). Among the chemical contaminants that can cause harm to the consumer present in milk, two classes of substance deserve to be highlighted due to the severity of the effects caused, mycotoxins and antimicrobial residues (GERMANO; GERMANO, 2019).

2.2.1 **Mycotoxins**

Mycotoxins are secondary metabolites produced by some filamentous fungi, observed in both plant and animal products, and can be detected at various levels of the food chain, their production being favoured by environmental stress conditions such as humidity and

temperature (DI CASTRO, 2015).
They are biologically active molecules with a low molecular weight that can contaminate various grains, causing economic impacts on agriculture and diseases in humans and animals after their metabolisation in the liver system (GONÇALVES, 2017).
Ingesting products contaminated by mycotoxins can lead to toxigenic diseases. Acute cases are associated with damage to the organs involved in purification and metabolisation, while prolonged exposure is associated with liver oncogenesis. There is a possibility of mutagenic alterations due to DNA damage and teratogenic alterations, resulting in childhood cancer during the embryonic period (BENKERROUM, 2020; IARC, 2002; KÓSZEGI; POÓR, 2016; KOVALSKY et al., 2016).

Aware of these risks, several authors have pointed out over the years the need for increased precaution and control of mycotoxins at a global level in order to minimise this intake (ALVITO et al., 2010; MEUCCI et al., 2009; TONON; SAVI; SCUSSEL, 2018).
There are currently more than four hundred known mycotoxins produced by hundreds of fungi. Mycotoxins can be divided into three large groups due to their characteristics: the aflatoxin group, produced by fungi of the genus Aspergillus; the ochratoxin group, produced by some species of the genera Aspergillus and Penicillium; and the fusariotoxin group, produced mainly by species of the genus Fusarium, with fumonisins, zearalenone and trichothecenes as the main representatives (SOUTO et al., 2017).
Although Normative Instruction (IN) 160 of 1 July 2022 does not set maximum tolerated limits (MRLs) for mycotoxins in raw milk, it does set MRLs for mycotoxins in fluid milk and various derivatives (BRASIL, 2022). The ingestion of raw milk by humans is not recommended due to the risk of pathogenic microorganisms from the incorrect milking process. Therefore, the MRLs for fluid milk are used as a parameter for the safety and quality of the raw material, since the high concentration of mycotoxins prevents it from being sent for processing and the preparation of derivatives (ADEYEYE, 2019).

2.2.1.1 Aflatoxins

Aflatoxins are produced by fungi of the genus Aspergillus, especially the species Aspergillus flavus and Aspergillus parasiticus, mainly in warmer and more humid climatic zones, and fungal proliferation can occur in products of plant origin due to failures during the production chain, such as harvesting, transport and storage (MASTANJEVIĆ et al., 2019).
There are four aflatoxins produced by the genus Aspergillus and they are classified according to the fluorescent colour they emit, with aflatoxins B1 and B2 having a blue colour and aflatoxins G1 and G2 having a green colour. Among the group, aflatoxin B1 (AFB1) is considered the main mycotoxin, both in terms of prevalence and toxicity (ADEYEYE, 2019; BENKERROUM, 2020).
AFB1 is mainly ingested by both animals and humans through the consumption of plant products contaminated with the toxin. During its passage through the hepatic system, AFB1 undergoes various biometabolisation reactions generating highly reactive intermediates capable of binding to the nitrogenous bases of the DNA molecule, forming DNA adducts and causing damage and teratogenic effects (GERMANO; GERMANO, 2019).

One of the possible biotransformation reactions for AFB1 is hydroxylation at carbon 10, transforming the compound into aflatoxin M1 (AFM1), increasing its hydrophilicity, which can be excreted in milk (VAZ et al., 2020). AFM1 is a thermostable molecule capable of resisting high heat treatments, however, there is the possibility of AFM1 irreversibly binding to milk casein, enhancing thermo-resistance, being able to resist pasteurisation and other processes for the production of dairy products, thus being transmitted to humans (GERMANO; GERMANO, 2019).

Brazilian legislation sets the MRL for aflatoxin M1 in fluid milk at 0.5µg/kg of the final product (BRASIL, 2022). This determination has been revised over the years, always with a view to adapting to stricter international standards such as the values described by the Codex Alimentarius in which concentrations of 0.05µg/kg or L of AFM1 in fluid milk are envisaged, thus demonstrating the increase in precaution and control of mycotoxins in the country (FAO, 2019; TONON; SAVI; SCUSSEL, 2018).

2.2.1.2 Zearalenone

Zearalenone (ZEA) is a mycotoxin produced by fungi of the Fusarium genus, mainly F. graminearum, F. culmorum, F. cerealis and other species. It can be formed in environments with temperatures between 20-25 °C and humidity above 20% (GIL- SERNA et al., 2014; KOVALSKY-PARIS et al., 2014; MOSTROM, 2012).

ZEA is metabolised by the intestinal cells of animals and can be biotransformed into four metabolites, the main ones being α-zearalenol (α-ZOL) and β- zearalenol (β-ZOL), which can be conjugated to glucoronic acid (UEBERSCHÄR et al., 2016; ZINEDINE et al., 2007).

The main biological effect of zearalenone and its derivatives is their oestrogenic activity, which can cause damage to the reproductive system of intoxicated animals. When evaluating biological activities, α-ZOL has greater oestrogenic activity than ZEA when metabolised by pigs. However, when metabolised by poultry, the metabolite produced in highest concentration is β-ZOL, which has lower oestrogenic activity (MOSTROM, 2012; RAI; DAS; TRIPATHI, 2019). Due to its toxicity, the presence of ZEA in food has been effectively monitored. Zearalenone can be present in grains and cereals, and can be present in both food and animal feed products (CHANG et al., 2020; GIL-SERNA et al., 2014; MALLY; SOLFRIZZO; DEGEN, 2016).

Zearalenone can be secreted into the milk when cows ingest the mycotoxin, but several studies have shown that the rate of passage is low, meaning that its residue in the milk of animals fed contaminated grain is of little clinical significance (COFFERY; CUMMINS; WARD, 2009; SIGNORINI et al., 2011; SMITH, 2006). However

However, due to the hydroxylation, glucoronidation and conjugation reactions involved in the biotransformation process, the metabolites produced can be present in the tissues and fluids of animals, and can be present in milk and pose health risks to consumers. Despite the risks associated with the presence of these metabolites, research into metabolites in milk has been neglected due to difficulties in detection and quantification (LIU; APPLEGATE, 2020; FALKAUSKAS et al., 2022).

2.2.2 **Antimicrobial residues**

Antimicrobials are compounds that reduce or inhibit microbial growth when used in optimal inhibitory concentrations. Antimicrobials can be divided into antibiotics and chemotherapies, with antibiotics being substances produced naturally by fungi, yeasts or other microorganisms and chemotherapies being chemical substances produced by synthesis (CARNEIRO, 2011). Antimicrobials can act in two ways on microorganisms: when the antimicrobial inhibits cell multiplication, but without causing lethal damage, it is called bacteriostatic, while when the lethal effect is evident, the antimicrobial is considered bactericidal. Some antimicrobials can have both bactericidal and bacteriostatic effects, depending only on the concentration and exposure time of the drug (SPINOSA; TÁRRAGA, 2019). Antimicrobials are important drugs used in human and veterinary medicine. In the field of veterinary medicine, antimicrobials are used for the therapeutic treatment of infections, prophylactic use to prevent diseases before or after exposure, and as feed additives to promote animal growth. They are widely used in all stages of production or the life cycle of animals, especially in mammary gland infections (mastitis) and respiratory tract diseases (BUZALSKI; REYBROECK, 1997; RONQUILLO; HERNANDEZ, 2017; SHAO et al., 2009). Parnham et al. (1998) stated that, theoretically, all routes of antimicrobial administration lead to the appearance of residues in food of animal origin. Therefore, the residues present in the animals' bodies can be transferred to the population through milk and its derivatives. The concentrations found in milk secretion can vary and can be attributed to various factors such as: the classes of antimicrobials, the type of antimicrobial formulation applied, variations between species, the pH difference between blood plasma and milk, the quantity of milk produced and the grace period (CAMPOS, 2001; NASCIMENTO; MAESTRO; SHAO et al., 2009). This presence is frequently reported in the literature (BANDO et al., 2009; BERENDSEN et al., 2010; LOPEZ et al., 2008; RODRIGUES; DALL'AGNOLB; BITTENCOURTA, 2012; TROMBETE; SANTOS; SOUZA, 2014).Among antimicrobial residues in food, the presence of residues of antimicrobials such as β-lactams, tetracyclines and aminoglycosides deserves to be highlighted. The residues generated by the administration of these drugs can be hazardous to consumers' health. They can cause allergic and toxic reactions in the short term in some sensitive individuals, and in the long term, they can result in chronic toxic effects or the development of antimicrobial-resistant bacteria in humans (AYTENFSU; MAMO; KEBEDE, 2016). The development of bacterial resistance due to the ingestion of food containing residues can influence the patient's treatment, leading to failures and recurrence, as well as the use of more toxic drugs for the patient (BANDEIRA et al., 2014; GASTALHO; SILVA; RAMOS, 2014; SILVA; TEJADA; TIMM, 2014).

The presence of antimicrobial residues in milk has detrimental effects on the production of dairy products, leading to failures in the production of fermented items (TREIBER; BERANEK-KANUER, 2021; VIRTO et al., 2022). These problems occur from their partial or total inhibition of the multiplication of lactic acid bacteria, resulting in a reduction in the concentration of acid formed by these bacteria. Consequently, this can lead to premature fermentation by pathogenic microorganisms and defects in the sensory characteristics of the final product (VIRTO et al., 2022; CHIESA et al., 2020).

In 2005, the Ministry of Health implemented the Veterinary Drug Residue Analysis

Programme in Brazil, which listed the groups of antimicrobials that should be monitored in dairy matrices. The report issued in 2009 highlighted the following groups of antimicrobials: β-lactams, tetracyclines, amphenicols, aminoglycosides and macrolides (BRASIL, 2009). Drug residue limits are determined by two programmes. The Ministry of Agriculture, Livestock and Supply operates through the National Plan for the Control of Residues and Contaminants (PNCR) (BRASIL, 1999), while ANVISA operates through the National Programme for the Analysis of Residues of Veterinary Medicines in Food Exposed to Consumption (PAMvet) (BRASIL, 2003a).

The processing industry routinely monitors the presence of antibiotic residues in milk. However, compliance checks are carried out seasonally, and the last report issued by PAMvet was in 2007. The samples are screened using rapid tests, such as the Simple and Aqueous Phase (SNAP) kits for β-lactams and tetracyclines and immunoenzymatic kits for aminoglycosides. After screening, the positive samples were submitted to specific confirmation and quantification tests for each antimicrobial (BRASIL, 2003a).

2.2.2.1 Azithromycin

Azithromycin is an antimicrobial belonging to the macrolide class and is one of the main representatives of this class due to its safety and broad spectrum of action. Macrolide antimicrobials are defined as having a macrocyclic lactone ring in their structure. Their mechanism of action is to inhibit protein synthesis in bacteria by binding to the 50S ribosomal subunit of the bacterial 70S ribosome, and they are indicated against bacterial infections caused by aerobic and facultative Gram-positive and Gram-negative bacteria (CORREA; FUKUSHIMA, 2020; GAUTRET et al, 2020NIH, 2022; SPINELLI et al., 2020).

Azithromycin has excellent pharmacokinetics, with tissue concentrations around 100 times higher than in plasma and 200 times higher in the intracellular space than in the extracellular space. It is metabolised very little and leaves no circulating active metabolites. The main adverse reactions caused by azithromycin are in the gastrointestinal tract and elevated levels of hepatic transaminases (RITTER et al., 2020).

The indiscriminate use of antimicrobials can also lead to another health risk, the development of resistance on the part of the target microorganisms. There are various types of mechanisms involved in antimicrobial resistance, in the case of macrolides, they can act directly against protein synthesis, inactivate the molecule or hinder the antimicrobial's access to the active site (BLAIR et al., 2015). The main mechanism of resistance to macrolides is the alteration of the target site on the bacterial ribosome. Activation of the erm gene demethylates the adenine nucleotide of the 23S subunit, altering its conformation 12, preventing the macrolide from binding to the ribosome and interrupting translocation of the peptide chain. Other mechanisms associated with macrolide resistance are those related to the cellular concentration of the drug, altering cell permeability or through efflux pumps preventing the antimicrobial from entering the cytoplasmic space (BLAIR, 2015; GRIFFITH, 2019; PARNHAM et al., 2014).

2.2.2.2 Streptomycin

Streptomycin belongs to the aminoglycoside class and was the first compound in this class to be discovered in 1944. The aminoglycoside class is used as broad-spectrum antimicrobials, inhibiting the growth of gram-positive and gram-negative aerobic bacteria, including most species of the Pseudomonas genus (ALFANDARI; CANNESSON, 2016).

Aminoglycosides act on protein synthesis in bacteria by binding to the 30S subunit of ribosomes and causing an irreversible effect on protein synthesis. The effect begins by entering the microorganism, mediated by active transport and passive diffusion, and its effectiveness depends on oxygen pressure and the existence of an electrochemical gradient on each side of the cell membrane (ALFANDARI; CANNESSON, 2016; MAIA; RATH; REYES, 2009).

The antimicrobial activity of aminoglycosides is inhibited in acidic pH environments and by bivalent cations, and their action is reduced in bronchial secretions, abscesses, tissue necrosis and large quantities of organic debris. Inactivation affects gentamicin and tobramycin the most and appears to be more intense in the presence of penicillins used to treat Pseudomonas. It is not recommended to administer aminoglycosides and β-lactams together (BRUNTON; CHABNER; KNOLLMAN, 2012; KATZUNG; TREVOR, 2017; RITTER et al., 2020).

Bacterial resistance to aminoglycosides appears after exposure to concentrations below the minimum inhibitory concentration (MIC) of the compounds. The main mechanism of resistance described is the synthesis of transferases from the group that catalyses the acetylation of amine functions and the transfer of phosphoryl or adenyl groups to the oxygen atoms of the hydroxyl radicals of each drug, since each aminoglycoside has a different sensitivity for each enzyme (BRUNTON; CHABNER; KNOLLMAN, 2012). Treatment protocols that utilise high doses to quickly reach the MIC/max ratio have been employed to prevent the development of this adaptive resistance by eliminating individuals with high MICs. (BRUNTON; CHABNER; KNOLLMAN, 2012; KATZUNG; TREVOR, 2017; RITTER et al., 2020) As streptomycin is also used as a veterinary drug, its residues can be found in various derivatives. Although the concentration of streptomycin in food has no acute toxic effect, several cases of allergic hypersensitivity associated with exposure to streptomycin have emerged over the years, which can produce severe rashes on the skin. Effective control of this antimicrobial in food is therefore necessary (MAIA; RATH; REYES, 2009).

2.2.2.3 Penicillin G

Penicillin was discovered in 1928 by Alexander Fleming when he observed plates inoculated with Staphylococcus aureus that had been accidentally contaminated by a fungus of the genus Penicillium showing bacterial inhibition due to some substance produced. It was only two decades later that the substance responsible for this inhibition began to be produced on a large scale and was known as penicillin G, playing an important role in the treatment of the wounded fighting in the Second World War (FLEMING, 1946; HILAL- DANDAN et al.,

2019).Penicillin G is an antimicrobial with a molecular formula belonging to the β-lactam group, characterised by the presence of a heterocyclic beta-lactam azetidinome ring in its structure, which is responsible for the activity of these drugs. The mechanism of action of penicillins occurs through the binding of the β-lactam ring with the penicillin-binding proteins (PBP) of the cell membrane of bacteria, preventing the formation of the peptidoglycan network, which results in defective cell walls that will lead to cell lysis (MARTIN; ALVAREZ-ALVAREZ; LIRAS, 2022). In the case of gram-negative bacteria, the permeability of the outer membrane makes it difficult for penicillin G to enter, which restricts its use to gram-positive bacteria and spirochetes (DOI, 2020; SIQUEIRA-BATISTA et al., 2021). Bacterial resistance to β-lactams can occur through the following mechanisms: inactivation of the drugs by β-lactamases, which is considered to be the main mechanism of resistance to penicillins; alteration of the transport system in the cell; modification of the permeability of the bacterial membrane; removal of the drug from the bacterial intracellular environment through an efflux pump; change of the binding site, through alteration of the active metabolic system for the drug; synthesis of alternative metabolic pathways (GARTLAN et al., 2022; MARTIN; ALVAREZ-ALVAREZ; LIRAS, 2022).

2.2.2.4 Tetracycline

The first member of the tetracycline family to be discovered was chlortetracycline by Benjamin Duggar in 1948 (DUGGAR, 1948). Chlortetracycline is a product of the fermentation of Streptomyces aureofaciens and has been the subject of several studies in the quest to obtain more effective synthetic derivatives, as well as research into new natural compounds from the family (PEREIRA-MAIA et al., 2010).

The mechanism of action of tetracyclines is related to the inhibition of protein synthesis in bacteria. The drug enters the bacterial cell by passive diffusion and binds to the 30S subunit of the ribosome, preventing the aminoacyl site of messenger RNA from binding to the 30S unit of ribosomal RNA (VICENTE; PÉREZ-TRALLERO, 2010).

Resistance is mainly produced by preventing the antimicrobial from binding to the target or by expelling the antimicrobial outside the cell by means of pumps. The determining factors in resistance are found in specific genes, located in mobile elements such as plasmids and transposons, and can be passed on to other bacteria (VICENTE; PÉREZ-TRALLERO, 2010).

The use of antimicrobials in dairy farming can lead to the presence of residues in milk, which can cause allergic reactions in consumers, induce resistance in bacteria and affect the development of cultures used to obtain fermented products. It is important to identify the levels of these contaminants in milk in order to maintain them as recommended by legislation, guaranteeing consumer safety and the quality of dairy products (PRADO; MACHINSKI JUNIOR, 2011).

2.3 METHODS FOR DETECTING CHEMICAL CONTAMINANTS

2.3.1 **Screening methods**

In order to carry out screening analyses on suspect samples, simple and easy-to-perform methodologies such as thin layer chromatography (DLC) and immunochemical techniques such as ELISA (Enzyme Linked Immuno- Sorbent Assay) and immunochromatographic tests (ICT) are used to detect mycotoxins. However Because the techniques are highly sensitive and can give a quantitative result for the detection of the analytes in question, they are commonly used as screening methodologies to qualitatively assess the presence or absence of mycotoxins (WHO, 2001).

The CCD technique consists of placing an aliquot of the sample to be evaluated on a chromatographic plate and, in parallel, an aliquot of the analyte standard previously solubilised in solvent. After the procedure, the plate is eluted with another solvent mixture and, after drying the plate, the plate is placed in an ultraviolet chamber in which it will be possible to evaluate the fluorescence of the runs and calculate the retention factor (RF) of the samples. The presence of a mycotoxin is confirmed if the sample has the same RF as the standard (SHUNDO; SABINO, 2006). The quantification of mycotoxins in the sample should be carried out using the recovery methodology described by Scott (1997).

Despite being a relatively easy technique to perform and routinely used, the number of publications using CCD has decreased over time, as information on its use and data has not been published (SHUNDO; SABINO, 2006; TRUCKNESS, 2001).

ELISA is a technique based on the principle of specific interaction between antibody and antigen. The technique consists of two steps, the first being the reaction between the antibody and the antigen and the second the revelation of the reaction by the enzymatic hydrolysis that occurs between the antigen-enzyme complex and the substrate (ANFOSSI; GIOVANNOLI; BAGGIANI, 2016).

The World Health Organisation recommends ELISA as a screening test for food of animal and plant origin (WHO, 2001). ELISA is a low-cost technique that is easy to perform and does not require specialised labour. Its use in food quality control analyses helps to optimise the aflatoxin detection process thanks to the method's sensitivity, specificity and speed (AFONSSI; GIOVANNOLI; BAGGIANI, 2016; WHO, 2001).

Zheng et al. (2006) described that ELISA is effective in detecting concentrations above 2.5ppb. However, with the high incidence of false-positive results, low reproducibility, variation of results from 30 to 300% and the possibility of false-negative results, it has been recommended that confirmatory techniques such as chromatography be used to confirm positive ELISA test results (AMARAL; MACHINSKI JUNIOR, 2006; HASSAM et al., 2021).In the case of dairy matrices, several studies have evaluated the efficiency of the ELISA technique in quantifying aflatoxin M1 and have found no significant variation between the added concentrations and the concentrations detected in the milk samples (OLIVEIRA; GERMANO, 1996; VAZ et al., 2020).In addition to immunoenzymatic methods, immunochemical-based tests in various formats are continually being developed with the aim of providing fast, portable and easy-to-operate systems (BELOGLAZOVA;

EREMIN, 2015; ZHANG et al., 2015). Despite presenting rapid results, immunochemical-based methods have a potential limitation in the simultaneous determination of different compounds and the detection of unknown toxins, as well as modified structures, due to the extreme selectivity of the molecular recognition mechanism. Strategies developed to overcome this problem have been the development of new tests that present the analytical platform in matrix format, in which several targets are detected separately in spatially defined zones (OSWALD et al., 2013; URUSOV et al., 2015).

ICT technology has played an important role in mycotoxin detection, being widely applied for the visual yes/no detection of mycotoxins and for their semi-quantification through the intensity of the result obtained (ANFOSSI; GIOVANNOLI; BAGGIANI, 2016; DZANTIEV; BYZOVA, 2014; MAJDINASABA et al., 2015; SUN et al., 2014).

The TIC approach has proved particularly suitable for this purpose, because it is precisely constructed as a strip along which the sample flows and encounters various bioreagents in different spatially confined zones, and several mycotoxins can be detected simultaneously (SONG et al., 2014; LI et al., 2013; ZANGHERI et al., 2015; WANG et al., 2013.

2.3.2 **High performance liquid chromatography**

High Performance Liquid Chromatography (HPLC) is a separation technique which, due to the possibility of altering the methodologies used, makes it possible to carry out quantitative determinations with good sensitivity, as well as the possibility of separating non-volatile and thermolabile species where gas chromatography cannot be used. As many compounds have the characteristics mentioned above, the field of application of HPLC is extremely vast.

HPLC features prominently in detections used by the pharmaceutical industry, environmental determinations and the monitoring of certain contaminants. In recent years, it has become one of the most widely used analytical methods for qualitative and quantitative purposes (SANKAR et al., 2020).

The aim of chromatography is to separate the various constituents of a mixture of substances individually in order to identify, quantify or obtain a pure substance. Separation takes place by passing the sample through a stationary phase via a solvent that acts as the mobile phase. After the sample is injected into the equipment, the components of the sample are distributed between the two phases according to their polarities and move more slowly than the mobile phase due to the force of attraction exerted by the stationary phase. The balance of these forces of attraction mediated by polarity determines the speed with which each component moves through the system, generating a specific retention time pattern for each substance, which can then be identified (DENOBILE; NASCIMENTO, 2004).

Despite all the qualities presented by the technique, the ability to identify

of the substances is limited. Errors can occur during the qualitative stage of the technique due to the chemical and structural characteristics of the substances analysed. Although the retention time is characteristic of a compound, several other compounds of similar polarity may have the same retention time under the chromatographic conditions used, even though they have different characteristics. This will result in co-elution between these compounds, with only one peak identified with greater intensity (FELTRIN et al., 2006).

Given the limitations of liquid chromatography, it has become necessary to use complementary techniques for identification. The use of the Mass Spectrometry (MS) technique has been coupled with liquid chromatography to make up for this deficiency, and has been used in routine studies to confirm target compounds rather than to determine the identity of unknown compounds of interest. In the LC/MS system, the sample separated by the HPLC is injected into the mass spectrometer using Ambient Pressure Ionisation (API) techniques, which generate a few ions to help determine the chemical structure of the analytes under investigation. The ions in turn pass through an analyser and the device emits peaks relative to the mass of the ions, allowing for more precise identification (LANÇAS, 2009).

2.3.3 **Spectrophotometry**

Spectroscopy is a branch of analytical chemistry that deals with the study of the interaction of electromagnetic radiation with matter. Traditionally, analyte interactions were between matter and electromagnetic radiation, but with spectroscopy, the interactions have been expanded to include interactions between matter and other forms of energy, such as particle beams like ions and electrons. These analytical methods are considered one of the most powerful tools available for studying the fundamental properties of materials and are also used to quantify a wide range of predominant chemical species in a given sample (KAFLÉ, 2017).

In spectroscopy, an analyst takes measurements of light (or light-induced charged particles) that are absorbed, emitted, reflected or scattered by a chemical analyte or material. Measurements are made by scanning a spectrum (point by point) or by simultaneously monitoring several positions on a spectrum and the measured data is then correlated to identify and quantify the chemical species present in that analyte (KAFLÉ, 2017).

Spectroscopic techniques are classified according to the type of radiation they employ and the way in which this radiation interacts with matter. These methods include those that use everything from radio waves to gamma rays and cause the nuclear spin to change in the nuclear nucleus. Electromagnetic (EM) radiation is made up of a stream of photons, travelling in a wave-like pattern at the speed of light. Each EM photon has a certain amount of energy. The type of radiation is defined by the amount of energy found in the photons and has both particle and wave properties, known as wave-particle duality, and comprises electric and magnetic fields. The EM spectrum comprises radiation, ranging from radio waves to gamma rays, and each radiation has a specific wavelength (KAFLÉ, 2017).

Ultraviolet and visible absorption spectrophotometry is a technique based on the attenuation of electromagnetic radiation by an absorbing substance [9]. This radiation has a spectral range o f approximately 190-800 nm, with the ultraviolet region being between 180-400nm and the visible range between 400-750nm. Attenuation results from reflection, scattering, absorption or interference. Accurate attenuation readings can be taken by recording absorbance alone. Within certain limits, absorbance is proportional to the concentration of the analyte to be determined and the distance of the light as it passes through the sample during irradiation, obeying Beer's Law. The relationship between absorbance and analyte concentration can be influenced by different factors such as the characteristics of the spectrophotometer, photodegradation of the molecules, the presence of scattering or absorption interferences in the sample, fluorescent compounds in the sample, interactions between the analyte and the solvent, pH and electrical current variations (PASSOS; SARAIVA, 2019). The radiation beam passing through the sample must pass through a detector in order to obtain the absorbance value of the sample. The function of a UV-Vis detector is to convert a light signal into an electrical signal and it must respond over a wide wavelength range, respond with high sensitivity and low noise, have a linear response range, have a fast response, allow miniaturisation and low sample consumption. As different substances absorb at different wavelengths, instruments must be able to control the wavelength of the incident electromagnetic radiation. In most instruments, this is done with a monochromator. In other instruments, this is done by using radiative filters or by using sources that emit radiation within a narrow wavelength range (KAFLÉ, 2017; PASSOS; SARAIVA, 2019).

2.4. ANTI-MYCOTOXIN ADDITIVES,

The main point to consider when controlling mycotoxins is preventing fungal growth, which consists of controlling their environment through good agricultural production and animal feed manufacturing practices. This process begins with soil preparation, seed selection, crop rotation, care in the field and with machinery, correct drying procedures, storage, transport and feed processing (TSIPLAKOU et al., 2014).

However, even with all the control measures, it is necessary to use substances capable of removing mycotoxins from food and from the animal's gastrointestinal tract after ingestion. Anti-mycotoxin additives, whether inorganic or organic, have been the solution found for application in various animal species as described by several authors (ADEGBEYE et al., 2020; BRETAS, 2018; CHUANG; HSIEH; LEE, 2020; COLOVIĆ et al., 2019; DANIELI; SCHOGOR, 2020; DALLMANN et al., 2021; REIS et al., 2020; ROSSI et al., 2013; OLIVER et al., 2020).

In response to this need, the Ministry of Agriculture, Livestock and Supply (MAPA) approved the creation of the Working Group on Mycotoxins in Products Intended for Animal Feed through Ministerial Order No. 13 of 24 May 2006, which has the following tasks: to assess the Brazilian situation regarding mycotoxin levels in products intended for animal feed with a focus on food safety; to define criteria for the control of mycotoxins of interest in products intended for animal feed; and to re-evaluate the use of mycotoxin adsorbents as an authorised additive in animal feed. One of the proposals made to MAPA by the group was to replace the term "mycotoxin adsorbent" by the general name "Anti-mycotoxin additives" (AAM), thus including products that, when added to animal feed, are capable of adsorbing, inactivating, neutralising or biotransforming mycotoxins. New products wishing to enter the national market must undergo individual in vivo and in vitro tests in order to guarantee the safety, capacity and harmlessness of the product in relation to animal health (BRASIL, 2006).

The classification of AAMs varies according to their mode of action: adsorbing agents and biotransforming agents. Adsorbent agents act by reducing exposure to mycotoxins, reducing their bioavailability, and can also be called binding agents and are mostly inorganic additives. On the other hand, biotransforming agents work by degrading mycotoxins into non-toxic metabolites using microorganisms and enzymes (BOUDERGUE et al., 2009). The differences inherent in each mycotoxin bioremoval mechanism will be further analysed in section 2.5.

2.4.1 **Inorganic adsorbents**

Inorganic adsorbents sequester toxins in the GIT, forming insoluble complexes through electrostatic bonds that are eliminated in the faeces. This reduces bioavailability and reduces the toxic effects of toxins. The majority are clays of volcanic origin such as bentonites and aluminosilicates, the latter being the class with the largest number of studies. Within this group, there are two important subclasses: phyllosilicates and tectosilicates. Phyllosilicates include bentonites, montmorillonites, smectites, kaolinites, etc.; and tectosilicates include zeolites (BOUDERGUE et al., 2009).

Montmorillonite is a phyllosilicate that was widely used in the early 1970s to reduce aflatoxins. Its structure is predominantly arranged in octahedral aluminium layers and tetrahedral silicon layers (ČOLOVIĆ et al., 2019; MASIMANGO; REMACLE; RAMAUT, 1978). Bentonite is generally an impure smectitic clay. Tectosilicates include important and highly studied zeolites. Zeolites consist of SiO4 and AlO4 tetrahedrons with a cage-like structure that is infinite in three dimensions. In these minerals, some tetravalent silicon is replaced by trivalent aluminium, which results in inorganic cations, such as sodium, calcium and potassium ions, that have no positive charge (ČOLOVIĆ et al., 2019; VILA-DONAT et al., 2018). The effectiveness of the adsorbent depends on the chemical structure of the adsorbent and the toxin. The main characteristic is the physical structure of the adsorbent, i.e. the total charge distribution, the pore dimensions and the available contact surface. However, this effectiveness also depends on characteristics such as the polarity, solubility, shape and charge distribution of the adsorbed substances. The stability of the toxin-adsorbent bond and effectiveness over a wide pH range are important criteria for evaluating possible mycotoxin binders because a product must be implemented throughout the gastrointestinal tract (de MIL et al., 2015; KELLER et al., 2015). One of the drawbacks of using inorganic substances to adsorb mycotoxins is their unspecific adsorption capacity, since they can sequester molecules of high nutritional value from the diet and certain veterinary medicines (ALAM; DENG, 2017; BARRIENTOS-VELÁZQUEZ et al, 2016; DEVREESE et al., 2013; ELLIOT; CONNOLLY; KOLAWOLE, 2020; FRANCISCATO et al., 2006; RALLA et al., 2010; WANG et al., 2018).

2.4.2 **Organic adsorbents**

In addition to inorganic adsorbents, various organic adsorbents have been used to control and mitigate mycotoxins, particularly charcoal and fermenting microorganisms. Increasingly, the removal of mycotoxins from food matrices has been based on technologies that utilise fermenting microorganisms. Control approaches have focused on biological detoxification, including biodegradation and bioadsorption between microorganisms and mycotoxins (LI et al., 2018).

When microorganisms are used as adsorbents, the amount of mycotoxins removed can vary depending on both the concentration of the toxin and the microorganism used. Different factors have influenced the adsorption capacity of biological agents, such as microbial concentrations, temperatures and pH values (PEREYRA et al., 2014).In addition to these factors, microorganism species are highly diverse in their cell wall composition and therefore in their adsorption capacity, showing effectiveness against a wide range of mycotoxins (LUO et al., 2019). This can be an environmentally friendly and effective strategy to ensure the safety of food contaminated with mycotoxins (WALTER; TUNDE; ISTVAN, 2015).

2.4.2.1 Activated carbon

Activated carbon is produced through the pyrolysis of organic materials and is considered to be a broad-spectrum adsorbent material with a large contact surface and, consequently, excellent adsorption capacity. Thanks to its porous structure, activated carbon is able to adsorb the main mycotoxins, including aflatoxins and zearalenone (JARD et al., 2011; LIU et al., 2022). Several studies have reported excellent adsorption rates using activated charcoal in animals. However, as it is a low-specificity adsorbent, its use has been reduced since activated charcoal can bind to essential nutrients, thus reducing their bioavailability and jeopardising the animal's nutrition. It is mainly used in cases of severe mycotoxin poisoning (HUWIG et al., 2001; HOLANDA; KIM, 2021.

2.4.2.2 Yeasts

Yeasts, whether viable or not, have a high adsorption capacity and are capable of reducing the bioavailability of various mycotoxins in food and feed. Yeast adsorption is carried out by surface adsorption of cell walls, and yeast species are highly diverse in cell wall composition and therefore vary in adsorption capacity (MCCORMICK, 2013; PFLIEGLER; PUSZTAHELYI; PÓCSI, 2015). Several yeast species have the ability to absorb mycotoxins, such as Saccharomyces cerevisiae, Candida tropicalis, Pichia pastoris and Phaffia rhodozyma (CECCHINI et al., 2019; LUO et al., 2016; PETERIA et al., 2007). S. cerevisiae stands out as a probiotic species and the yeast closest to humans, playing an important role in the adsorption of mycotoxins. Adsorbents based on yeast cell wall (PCL), such as S. cerevisiae, have esterified glucomannans and are capable of efficiently binding to various mycotoxins, such as aflatoxins, fumonisins and zearalenone (ARMANDO et al, 2012; CECCHINI et al., 2019; GONÇALVES et al., 2015; PETRUZZI et al., 2016; PIZZOLITTO; SALVANO; DALCERO, 2012; POLONI et al., 2015; ZOU et al., 2015). The constituents of PCL from S. cerevisiae are mannans, proteins and β-glucans. The mannans are responsible for the prebiotic property of the wall, modulating the intestinal microflora and improving intestinal integrity, reducing turnover and modulating the system. in the intestinal lumen. The β-glucan portion has the ability to stimulate the immune system's white blood cells, but its function of interest in this study is to adsorb mycotoxins ingested in the diet and help eliminate them (GRAHAM; MCCRACKEN, 2005; KLIS et al., 2006; YIANNIKOURIS et al., 2003).However, the results in the literature are diverse, showing that more factors probably interfere with adsorption efficiency, such as differences between yeast strains, environmental conditions, pH of the medium, cultivation medium and the amount of esterified glucomannan in the composition of the additive. PCL-based products, even though they have a similar basic constitution, are different and can vary in terms of their adsorption capacity and must be tested individually to be considered as AAM, as described in various studies (KELLER et al, 2012; KELLER et al., 2015; MALLMANN; DILKIN, 2007; YIANNIKOURIS et al., 2003; YIANNIKOURIS et al., 2004a; YIANNIKOURIS et al., 2004b).

2.4.2.3 Lactic acid bacteria

The group of lactic acid bacteria (LAB) is made up of Gram-positive bacteria, of varied morphology, mesophilic, generally immobile and non-spore forming, of different genera, facultative anaerobic or microaerophilic. The main characteristic of this group is their ability to ferment glucose, obtained from the hydrolysis of lactose, producing lactic acid. They are also capable of producing various antimicrobial factors, such as organic acids, hydrogen peroxide, nisins and bacteriocins. This group includes the genera Pediococcus, Streptococcus, Lactococcus, Leuconostoc, Lactobacillus and Bifidobacterium (FORSYTHE, 2002; OLIVEIRA, 2009; WALSTRA; WOUTERS; GEURTS, 2006).

LAB are often used to obtain fermented foods, meats, vegetables, drinks, fruits and as probiotics, and can also be found as constituents of the respiratory and gastrointestinal tracts, as well as the cavities of humans and animals (FORSYTHE, 2002).

Industrially, fermentation by lactic acid bacteria also helps to preserve food by producing antibacterial agents and acidifying the environment. However, the presence of lactic acid bacteria can also be harmful when it causes the pH to drop, as a result of the accumulation of lactic acid produced, leading to the precipitation of casein in raw milk (OLIVEIRA, 2009; ORDÓÑEZ, 2005).Over the years, various studies have shown the benefits of bacteriocins produced by BAL. These thermostable substances are antimicrobial peptides synthesised in bacterial ribosomes and have bacteriostatic and bacteriocidal activity depending on their type and concentration (DELAVENNE et al., 2012; SHARMA et al., 2021; ZHENG et al., 2014).

The main characteristic of choice for the use of a bacteriocin is the ability to exert antimicrobial activity against different bacteria, fungi, parasites, viruses and in natural resistant structures, such as bacterial biofilms (EYANG et al., 2014; GRAHAM et al., 2017; PANG et al., 2022), especially against multi-resistant bacterial infections (LANGDON; CROOK; DANTAS, 2016; SOLTANI et al., 2021). It is also important to emphasise the antimicrobial effect generated through the production of organic acids, causing a reduction in the pH of the matrix inhibiting the growth of susceptible microorganisms (CIZEIKIENE et al., 2013).According to the United Nations, probiotics are defined as live microorganisms that, when administered in adequate quantities, beneficially affect the health of the host (WAN; FORSYTHE; EL-NAZAMI, 2019; WHO/FAO, 2002). Several benefits are associated with the intake of probiotics by humans, including the prevention of infections, gastrointestinal disorders, allergic processes, carcinogenesis and tumour growth (MARTINEZ, BEDANI; SAAD, 2015; MORENO DE LE BLANC; LE BLANC, 2014). Recently, several studies on the use of probiotics have shown significant effects in preventing urogenital, respiratory and skin infections, as well as helping with glycaemic control and reducing risk factors for metabolic syndrome (BORDALO TONUCCI et al., 2017; BUSTAMANTE et al., 2020; XAVIER-SANTOS et al., 2020). To be considered a probiotic, the microorganisms must be able to survive gastric and enteric conditions and adhere well to the intestinal mucosa, since most of the effects occur in this region. They must multiply until they reach the concentration required for the product to be considered probiotic (DOS SANTOS et al., 2015). In Brazil, current legislation for products with probiotic claims states that the appropriate concentration of microorganisms must be specified in order to exert their probiotic effect. However, the literature has reported concentrations of between 10^6 - 10^9 CFU/g or mL for the product to be classified as probiotic (FAO, 2002; SHAH, 2007; GRANATO et al., 2010). It is important to emphasise that guaranteeing the conditions for a product to be considered probiotic is one of the industry's challenges, since any alteration to the product can interfere with the concentration. In addition, the intrinsic variations of each microorganism can end up interfering with the maintenance of the final concentration (XAVIER-SANTOS et al., 2020).

One of the main objectives of biotechnological processes applied to probiotics is to improve the technological characteristics of these microorganisms, improving the utilisation of the nutrients available in the food, improving the production of metabolites of interest for the characteristics of the product, increasing the survival and viability of the probiotic in the product. One way of obtaining these optimised strains is through strain engineering using genes from other species of lactic acid bacteria, genes from other groups of bacteria, or even gene silencing (DOUILLARD; DE VOS, 2019).

Because of their inherent benefits and functionalities, the number of people interested in consuming probiotic products is growing. The most common genera of LAB currently used as probiotics are Lactobacillus and Bifidobacterium (SNIFFEN et al., 2018; TYUTKOV et al., 2022).In addition to their fermentative, probiotic and microbial growth inhibition characteristics, several studies point to the ability of lactic acid bacteria to adsorb, inhibit and biometabolise toxins present in the dairy matrix, such as mycotoxins produced by the secondary metabolism of filamentous fungi (FUCHS, 2008; MARTINEZ, 2019).

Due to certain characteristics, specific strains of probiotic bacteria have been indicated as effective in removing contaminants such as heavy metals, pesticides and mycotoxins, and can act as promising bioagents for food safety (CHIOCCHETTI et al., 2019; ZOGHIA; KHOSRAVI-DARANI; SOHRABVANDIB, 2014).

One of the main mechanisms involved in this removal of contaminants is the cell wall. The cell wall of LAB is mainly composed of a layer of peptidoglycans embedded with teichoic and lipoteichoic acids and polysaccharides (HATHOUT; ALY, 2014). The mechanisms of contaminant removal by probiotics, whether lactic acid bacteria or yeasts, will be discussed in the next section.

2.5 BIOTECHNOLOGICAL APPLICATION

According to Article 2 of the Convention on Biological Diversity, the concept of biotechnology is defined as any technological application that uses biological systems, living organisms or derivatives thereof, to obtain or modify products and processes for a specific use (UN, 1992). It was subsequently ratified by 168 countries and accepted by the Food and Agriculture Organisation (FAO). Agriculture Organisation (FAO) and the World Health

Organisation (WHO), as "any technological application that uses biological systems, living organisms or their derivatives, to create or modify products and processes for specific uses" (FERRO, 2010). Among biotechnological processes, those aimed at improving microorganism cultures for use in food processing are part of what is known as traditional biotechnology and include classic genetic improvement methods such as mutagenesis and conjugation. Fermentative bioprocesses stand out as the main biotechnological application of microorganisms in food processing (FAO, 2011; NISSAR et al, 2017).Fermentative bioprocesses have been widely applied due to the probiotic characteristics of microorganisms, especially the production of beneficial compounds for the product and the consumer. Applications include the removal of toxins that may be present in food, such as bioadsorption and biodegradation.

2.5.1 **Bioadsorption**

Adsorption is a mass transfer operation which studies the ability of certain solids to concentrate certain substances on their surface in liquid or gaseous fluids, making it possible to separate the components of these fluids. Since the adsorbed components are concentrated on the external surface, the greater the external surface per unit of solid mass, the more favourable the adsorption. Because of these characteristics, most adsorbent materials are solids with porous particles. The species that accumulates at the interface of the material is usually called the adsorbate or adsorbate, and the solid surface on which the adsorbate accumulates is called the adsorbent or adsorbent (RUTHVEN, 1984).

Adsorption separation processes are based on three distinct mechanisms: the steric mechanism, equilibrium mechanisms and kinetic mechanisms. For the steric mechanism, the pores of the adsorbent material have characteristic dimensions, which allow certain molecules to enter, excluding the others. The equilibrium mechanisms are based on the ability of different solids to accommodate different adsorbate species, which are preferentially adsorbed to other compounds. The kinetic mechanism is based on the different diffusivities of the various species in the adsorbent pores (RUTHVEN, 1984). In parallel, bioadsorption technology uses the cell surfaces of inactivated microorganisms to bind various mycotoxins in food and feed, reducing their bioavailability in the product. The ability of some bacteria and yeasts to bind mycotoxins in food has already been demonstrated by SHETTY AND JESPERSEN (2006) and has been the basis for various works and technological innovations in the area of mycotoxin prevention and control, being an effective strategy with a beneficial environmental impact (WALTER; TUNDE; ISTVAN, 2015).

Studies using Planococcus spp. have demonstrated zearalenone removal in several of the experimental conditions, with excellent results at 30 °C, pH 4.5 using a concentration greater than 5×10^7 CFU/mL (LU; LIANG; CHEN, 2011). Another study evaluating the removal of patulin from apple juice using inactivated yeasts found that adsorption increased using a longer incubation time (30 hours) at pH 5 (GUO et al., 2012a).

According to Luo et al. (2015), the removal of patulin by active and inactive yeast cells showed a significantly similar adsorption capacity of the same strain, and the adsorption rate of patulin from yeast cells decreased sharply when they were damaged or lost their cell walls, which indicated that this behaviour was more likely caused by adsorption on the cell surface than by complex enzymatic reactions.

The use of lactic acid bacteria is also effective in reducing patulin concentrations in fruit juices, showing excellent results using both viable and inactivated cells, demonstrating that inactivated cells can be used effectively without altering the sensory characteristics of the product (LUO et al., 2016). According to Hatab and collaborades (2012a), the use of inactivated LAB demonstrated an adsorption capacity against patulin comparable to viable cells in aqueous solution.

During microbial adsorption, the amount of toxins removed can vary due to reaction conditions such as pH, incubation time and temperature, as well as the concentration of the toxin and the microorganisms in the reaction medium (GUO et al., 2012; LU; LIANG; CHEN, 2011; PEREYRA et al., 2014; WALTER; TUNDE; ISTVAN, 2015).

Microorganisms also have another variable, which is the high diversity of their cell wall composition, which will have different hydrophobic characteristics. Mycotoxin adsorption occurs through hydrophobic and electrostatic interactions, such as weak hydrogen bonds and van der Waals bonds, between the metabolite and the binding sites formed by the constituents of the yeast cells. Additionally, the higher the number of membrane constituents, the greater the number of binding sites and the adsorptive capacity (ARMANDO et al., 2012; JOUANY; YIANNIKOURIS; BERTIN, 2005; PEREYRA et al., 2018; PFLIEGLER; PUSZTAHELYI; PÓCSI, 2015).

The viability of the microorganisms is also capable of influencing the removal of microorganisms since the removal of mycotoxins can occur through other mechanisms, such as biotransformation of the compound and degradation by the enzyme complex (BANGAR et al., 2021).

The mechanisms involved in adsorption phenomena are varied due to the variety of species used as biological adsorbents (WANG et al., 2015). Yeasts with an integrated cell wall have a more effective adsorptive capacity than other species, indicating that the structural integrity of the wall is important for the adsorptive effect (ARMANDO et al., 2012; GUO et al., 2012b).

Based on this information, several theories have been developed to explain adsorption using yeasts, including the theory of cell morphology and physical structure; the theory of membrane chemical components; and the theory relating the interactions between membrane components and mycotoxins (LUO et al., 2020).

The chemical components of the cell membrane play a fundamental role in the formation of the mycotoxin-yeast complex. The yeast cell wall is composed of polysaccharides with an outer layer composed of heavily glycosylated mannoproteins and an inner layer composed of β-1,6- and β-1,3-glucan chains linked to chitin and mannoproteins (GONÇALVES et al., 2015; PETRUZZI et al., 2016).

Studies show that insoluble alkaline β-1,3-glucans form a complex structure with a large number of binding sites for ZEN and soluble alkaline β-1,6-glucans enhance this binding (AAZAMI et al., 2018; HAMZA et al., 2016).

The solubilisation of β-glucans in different pH conditions induces changes in structure and conformational energy to create new interactions. The variation in pH alters the flexibility of these interactions, leading to greater flexibility in neutral conditions than that observed in more acidic conditions. The decrease in flexibility at pH 3.0 causes an increase in adsorption values for ZEN when compared to adsorption under neutral and alkaline conditions

(AAZAMI et al., 2018; MARTÍNEZ et al., 2023).
Similarly, the adsorptive capacity of LAB is also realised through the constituent components of the cell wall, particularly peptidoglycans and membrane polysaccharides. However, the structure of peptidoglycans is not the main factor involved in adsorptive phenomena; in heat-inactivated bacteria, the adsorption of aflatoxins was more effective due to the mycotoxin's binding to carbohydrates and protein components (BOVO et al., 2013; WANG et al., 2015b).
According to Niderkorn et al. (2009), the binding efficiency of different LAB species seems to be more closely related to the amino acid sequence of the peptide bridges in the peptidoglycan.
Zoghia et al. (2014) demonstrated that the cell wall of LAB contains many negatively charged functional groups that can facilitate binding capacity due to the presence of S-layer proteins
Studies investigating the binding of AFB1 to bacteria of the genus Lactobacillus and Propionibacterium in vitro have shown that Lacticaseibacillus rhamnosus GG and L. rhamnosus LC705 can adsorb around 80 per cent of AFB1 (5 g/mL) in one hour at a concentration of 10^{10} CFU/mL (HASKARD et al., 2001; ZHANG et al., 2023). After heat inactivation and acid treatment, the two Lactobacillus strains proved to be more effective in reducing AFB1 and exhibiting a better binding capacity for ZEN and its derivative - ZOL (EL-NAZAMI et al., 2002; ZHANG et al., 2023).
Despite the difference in effectiveness compared to inorganic adsorbents, the use of live microorganisms has been the focus of study for the mitigation of mycotoxins in various species due to the probiotic effect and the biodegradation of mycotoxins associated with the ingestion of microorganisms (KELLER et al., 2015).

2.5.2 **Biodegradation**

Different approaches have been investigated in order to eliminate the presence of post-harvest mycotoxins. Since traditional physical and chemical strategies have limited effectiveness and can cause losses in nutritional value, present biosafety risks and high potential costs, the removal of mycotoxins has increasingly been based on their interactions with microorganisms, with emphasis on the aforementioned bioadsorption and biodegradation (JI et al., 2016; LI et al., 2018).
Biodegradation technology is based on the use of microorganisms and/or enzymes to degrade mycotoxins into inert or less toxic compounds, depending on the metabolic pathway of the living microorganisms (LI et al., 2018; PETRUZZI et al., 2014; SARLAK et al., 2017). In contrast to studies involving adsorption mechanisms, research into the biodegradation of mycotoxins by probiotics is very limited (LIU; XIE; WEI, 2022). Yeast biodegradation mechanisms can occur through isolated enzymes or the presence of the yeast itself, which can be degraded either by enzymes released into the environment or by biometabolisation processes. As demonstrated by Cao et al. (2011) who evaluated the degradation of aflatoxin B1 against an oxidative enzyme from the fungus Armillariella tabescens and proposed that the inactivation mechanism occurred through the cleavage of the bis-furan ring of the aflatoxin molecule.In another study using S. cerevisiae, Keller et al. (2015) showed that it took 18 hours to reduce around 90 per cent of zearalenone into its metabolites, α-ZOL and β-ZOL,

reducing its bioavailability. Although it has an excellent conversion rate, its application in food is unfeasible due to the alterations that can occur as a result of exposure time. However, they can be used as an additive in animal nutrition, increasing productivity and health parameters, as well as minimising the bioavailability of mycotoxins present in the diet (GONZÁLEZ-PEREYRA et al., 2014).

When evaluating the biodegradation of bacteria, various factors can alter the degradation processes, such as incubation period, reaction medium, bacterial species, cell concentration and pH (NDIAYE et al., 2022).

Many species of bacteria, such as BAL, have the ability to degrade mycotoxins into other metabolites and, like bioadsorption, the process of mycotoxin degradation by bacteria varies between each species (SADIQ et al., 2019). The genetic variations of each strain of the same species are also capable of influencing the speed of degradation. As demonstrated by El-Nezami et al. (1998) who evaluated the removal of AFB1 in vitro using five LAB, two of which were strains of Lacticaseibacillus rhamnosus. The result obtained showed a greater reductive effect of the L. rhamnosus strains and a difference between the two strains.

It is important to emphasise that biodegradation methods can pose potential risks. Most studies do not evaluate the possibility of new toxic by-products being formed during the decomposition process. For large-scale biotechnological use, applicable methods must be developed to monitor the potential hazardous metabolites and biological effects in the biodegradation of mycotoxins (LIU; XIE; WEI, 2022).

2.5.3 **Lactose metabolisation and lactic acid production**

Lactose is the main carbohydrate in milk, with an average concentration of 4.8 per cent (SURI et al., 2019). According to Normative Instruction No. 76 of 2018, which determines According to the Technical Regulation on the Identity and Quality of Raw Milk, the lactose content in raw milk must be at least 4.3 per cent (4.3g/100mL) to be considered as such (BRASIL, 2018).

Lactose is a disaccharide made up of glucose and galactose. It consists of a molecule of alpha or beta-galactose and a molecule of beta-glucose, acting as a regulator of osmotic control. In milk, it is still possible to find the molecules that make up lactose in their hydrolysed form free in milk in lower concentrations of glucose and galactose (SCHULZ; RIZVI, 2018; SILVA; VIDAL; NETTO, 2018).

Among lactose-related problems, lactose intolerance is the most prevalent among humans. Lactose intolerance can be defined as the inability to digest lactose due to a deficiency or absence of the enzyme lactase, which is responsible for hydrolysing it into its constituent monosaccharides (BATISTA et al., 2016).

Non-hydrolysed lactose increases the osmolarity of the small intestine, where it attracts water and electrolytes to the intestinal mucosa, resulting in diarrhoea. Intestinal dilation ends up contributing to the malabsorption of lactose, which serves as a substrate for the intestinal microbiota and is fermented, causing malaise, nausea, vomiting, diarrhoea and flatulence (GALVÃO, 2012).

Therefore, the way for people with this condition to consume dairy products is to remove this carbohydrate. Lactose is removed from milk through hydrolysis reactions, which can be acidic or enzymatic (SCHULZ; RIZVI, 2018).The acid hydrolysis reaction is fast and involves dilute solutions of strong acids and high pH and temperature conditions, 1.0-2.0 and 100-150°C respectively. Despite the advantage of speed, its application in the dairy products industry is restricted due to changes in the colour and flavour of the products, as well as nutritional losses due to the denaturation of milk proteins (SCHULZ; RIZVI, 2018).

Enzymatic hydrolysis requires no prior treatment and can be applied directly to the milk or whey. Hydrolysis is catalysed by the enzyme lactase (β-galactosidase) and takes place under mild pH and temperature conditions, reducing the occurrence of matrix alteration, the formation of unwanted by-products and wear and tear on equipment. The products obtained in this way preserve their properties, increasing their relative sweetening power (SCHULZ; RIZVI, 2018).

The enzymatic hydrolysis of lactose can also be done through the fermentative mechanism of obtaining energy from anaerobic or microaerophilic microorganisms. It is worth emphasising that the production of commercial products obtained through microbial fermentation has Green technology is gaining prominence due to the increasing demand for energy and the environmental problems caused by burning fossil fuels. The development of new techniques to produce green technology products in a sustainable way is essential (HASSAN et al., 2020; SRIVASTAVA et al., 2020).

The main product formed through the fermentation of lactose is lactic acid, which is of industrial interest due to its use in the manufacture of food and beverages, medicines, cosmetics and textiles (RAWOOF et al., 2021). Lactic acid (CH3- CHOHCOOH) has two enantiomeric forms, L (+) lactic acid and D (-) lactic acid, each with their respective industrial applications. Lactic acid with high optical purity is more valuable than the racemic form, enabling broader industrial applications; in the food and beverage industry, for example, the levorotatory form is more widely used due to its particularities (QIU et al., 2019).

The fermentation of lactose to lactic acid begins with the enzymatic hydrolysis of lactose catalysed by lactase enzymes of microbial origin. After this process, the microorganism assimilates the constituent molecules, glucose and galactose, to produce energy (TORTORA; FUNK; CASE, 2017).

The glucose will then undergo the process of glycolysis, in which it will be oxidised, giving rise to two molecules of pyruvic acid, producing ATP and NADH in the process. The pyruvic acid produced will be reduced by molecules of NADH and form lactic acid at the end of the fermentation process in the proportion of two molecules of lactic acid for each molecule of glucose consumed (TORTORA; FUNK; CASE, 2017).

Due to the genetic variety of fermenting microorganisms, various products and by-products can be obtained through microbial fermentation, being classified as homofermentative and heterofermentative. In homofermentation, the production of lactic acid is around 85% of the glucose consumed, while in heterofermentation the proportion is around 50% and other compounds such as ethanol and carbon dioxide are also produced (RAWOOF et al., 2021).

2.5.4 **Factors influencing microbial growth**

As described above, there are a number of intrinsic and extrinsic factors involved in microbial growth. Among these factors of interest for biotechnological applications, we can highlight: composition of the medium/food matrix gas atmosphere, temperature, stress, pH, cell integrity, the presence of other microorganisms, and the presence of antimicrobial substances (ROLFE; DARYAEI, 2020).

Microorganisms need nutrients to grow and remain viable in the environment. Nutrients are used as sources of energy and raw material for microbial development, the main nutrients being carbon, nitrogen, minerals, vitamins and other growth factors (HAMAD, 2012).

The ability of a matrix to support microbial growth depends on its content of these nutrients. Almost all microorganisms use simple sugars, present in most foods, such as glucose, fructose, sucrose and maltose, as energy and carbon sources. Amino acids, alcohols and free fatty acids are also used by many microorganisms as energy and carbon sources (HAMAD, 2012; ROLFE; DARYAEI, 2020).

In studies with yeasts, Pereyra et al. (2018) demonstrated that the ability

The adsorptive effect of S. cerevisae is influenced by the composition of the microorganisms' culture medium, and this effect is observed through changes in the thickness of the cell wall as a result of variations in the nutrients in the medium.

In another study, Lima et al. (2009) evaluated the difference in the growth of lactic acid bacteria in different culture media and were able to demonstrate the differences between each cell growth in different temperature and atmosphere conditions during incubation.

Microorganisms are subject to various factors that can alter their physical and chemical environment, which can cause cellular stress at various stages of the production chain. Stress situations, when applied within a certain range, allow cells to develop acquired resistance mechanisms to survive the situations they are subjected to, as described below (ROLFE; DARYAEI, 2020).

Among the mechanisms of resistance to acidic stresses are acid resistance, when subjected to a mild acidic environment (pH 5.0-5.8); acid tolerance response (ATR), when exposed to pH between 2.4 and 4.0; and acid shock response (ASR), when microorganisms are exposed to extremely acidic environments before undergoing prior adaptation at moderate pH (RAY; BHUNIA, 2013).

Microorganisms grow in a wide range of pH values extending from below pH 1 to around pH 11. Fungi have the widest range, followed by yeasts, while bacteria have a narrower pH growth range (JAY et al., 2000). In general, yeasts and moulds grow best in acidic environments, although some bacteria can also grow (JAY, 2000; LUND; EKLUND, 2000). Bacteria grow in a pH range of 4.5 to 9.0, while yeasts and filamentous fungi grow in a pH range of 2.0 to 10.0. The optimal growth of most microorganisms is observed at neutral pH levels (6.6-7.5), with a small number being able to grow below a pH of 4.0 (JAY et al. 2000).

Fermentation with the final production of organic acids can lower pH levels and, in turn, cause growth restrictions and/or the death of various microorganisms. The inhibition of pathogenic microorganisms due to the reduction in pH is one of the main factors considered when making fermented products (HAMAD, 2012; LEVINE; FELLERS, 1940).

The presence of several microorganisms in the matrix can result in competition for the same nutrients, which can cause a stress situation. As a result, the production of toxic by-products such as bacteriocins, organic acids, harmful metabolites and antimicrobials can be produced by various types of microorganisms with a view to the survival of the species by inhibiting the growth of competing species (BYAKIKA et al., 2019; LI et al., 2021; LIAO et al., 2020; REN et al., 2019).

Metabolites with antimicrobial activity are small peptides produced by organisms and play an important role in the body's innate immunity (BAHAR; REN, 2013). As well as being effective against pathogens, they also act to regulate the host's autoimmune system, serving as the basis for applications in the prevention and treatment of animal and plant-related diseases, in the development of new drugs and in the field of biological detoxification (FRY, 2018).

Bacteria of the genus Lactobacillus spp. have high antifungal activity due to the production of organic acids, primarily lactic acid and lower concentrations of phenylactic acid, hydroxyphenylactic acid, indole lactic acid, 2-butyl-4-hexyloctahydro-1H-indene, oleic acid, palmitic acid, linoleic acid and 2,4-di-tert-butylphenol (GUIMARÃES et al., 2018; RUSSO et al., 2017; SANGMANEE; HONGPATTARAKERE, 2014).

Due to the variety of biosynthetic routes, yeasts have been increasingly used to produce different bioactive substances (SIDDIQUI et al., 2012). Several studies evaluating the biocontrol activity of yeasts have proved to be very effective in controlling growth, gene expression and aflatoxin production in various species of the Aspergillus sp. genus, such as the production of 2-phenylethanol, β-1,3-glucanase and exo-quitinase by Pichia anomala (HUA et al., 2014; TAYEL et al., 2013), isoamyl acetate and isoamyl alcohol by Candida maltosa (ANDO et al., 2012) and 4-hydroxyphenethyl alcohol, 4,4- dimethyloxazole and 1,2-benzenedicarboxylic acid diocyl ester by S. cerevisae (ABDEL-KAREEM; RASMEY; ZOHRI, 2019).

3. OBJECTIVES

3.1 GENERAL OBJECTIVE

Evaluate the mitigation potential of probiotic microorganisms against mycotoxins and antimicrobial residues, assessing the adsorptive and biodegradation potential of microorganisms in vitro and in raw milk.

3.2 SPECIFIC OBJECTIVES

1. Establishment of methodologies for making inoculums with standardised concentrations
2. Evaluate the adsorptive effects of wild yeast strains against zearalenone and aflatoxin
3. Construction of microbial growth kinetics curves for the isolated strains in order to evaluate the generation time and the maximum specific growth rate.
4. Evaluation of resistance to certain antimicrobial agents in order to trace the resistance profile of the isolated bacteria.
5. Evaluate the lactose consumption rate of the isolated strains and assess their effectiveness for use in the production of dairy products
6. Evaluate the adsorptive and biotransformation effects against mycotoxins and antimicrobials;

4 MATERIAL AND METHODS

4.1 MATERIAL

The equipment and most of the consumables needed to carry out the research were available at the Laboratory for the Microbiological Control of Products of Animal Origin of the Department of Food Technology and the Laboratory for the Physical-Chemical Control of Products of Animal Origin of the Faculty of Veterinary Medicine of the Fluminense Federal University and the mycology and mycotoxicology laboratory of the Federal University of Minas Gerais (LAMICO-UFMG).Three strains of S. cerevisiae isolated from bovine fodder, coded as S. cerevisiae LL74, LL08 and LL83, were used for the yeast adsorption tests. The strains were deposited in the culture collection of the Mycology and Mycotoxins Laboratory belonging to the collection centre of the Federal University of Minas Gerais (LAMICO - UFMG) and the National University of Río Cuarto (Córdoba, Argentina). A commercial strain of S. cerevisiae (coded as CS) was used as a control, at the recommended dose of 2.0mg/mL (2mg/g per feed, with a cell count ≥ 107 UFC/g).Lactobacillus strains used as probiotics for human use were used. The strains were purchased from a commercial compounding pharmacy. The packaging containing the freeze-dried strains was opened when the capsules were prepared. The following strains were used: Bifdobacterium lactis Bifido LYO® (Fermentech, Tatuapé, Brazil), LACT GB® (Gabbia Biotecnologia e Desenvolvimento, Barra Velha, Brazil); Lactobacillus acidophilus ACID GB® (Gabbia Biotecnologia e Desenvolvimento, Barra velha, Brazil), LA-5® (Chr. Hansen, Hørsholm, Denmark), acidophilus LYO® (Fermentech, Brazil); Lacticaseibacillus delbrueckii subsp. bulgaricus LB-G40® (BioGrowing Co., Shanghai, China), LB 340 LYO® (Fermentech, Brazil), BULG GB® (Gabbia Biotecnologia e Desenvolvimento, Barra Velha, Brazil); Lacticaseibacillus paracasei NTU 101® (Centro Sperimentale Del Latte, Zelo Buon Persico, Italy), L. CASEI 431® (Chr. Hansen, Hørsholm, Denmark), CAS GB® (Gabbia Biotecnologia e Desenvolvimento, Barra Velha, Brazil); Lactobacillus Gasseri LG08® (Jiangsu Wecare Biotechnology Co, Jiangsu, China), GASS GB® (Gabbia Biotecnologia e Desenvolvimento, Barra Velha, Brazil); Limosillactibacillus reuteri RC-14® (Chr. Hansen, Hørsholm, Denmark); Lactobacillus rhamnosus CRL 1505® (Centro Sperimentale Del Latte, Zelo Buon Persico, Italy), GR-1® (Chr. Hansen, Hørsholm, Denmark), LGG® (Chr. Hansen, Hørsholm, Denmark),Rhamnosus LYO® (Fermentech, Tatuapé, Brazil); Streptococcus thermophilus TH-4® (Chr. Hansen, Hørsholm, Denmark), TA 40 LYO® (Chr. Hansen, Hørsholm, Denmark).The culture media used were MRS broth (de Man, Rugosa and Sharpe) (KASVI® , São José dos Pinhais, Brazil) for the growth of lactic acid bacteria and YPD broth (1% yeast extract, 2% peptone and 2% glucose) for the cultivation and growth of S. cerevisiae strains.The following standards were used in this work: aflatoxin B1 (TAS-M11LA1-10®Trilogy Analytical Laboratory, Washington, USA), aflatoxin M1 (TAS-M15® Trilogy Analytical Laboratory, Washington, USA) and zearalenone (TAS-M17LM2-10® Trilogy Analytical Laboratory, Washington, USA) and the antimicrobial standards: Azithromycin, Streptomycin, Tetracycline and Penicillin G (Sigma-Aldrich, St. Louis, USA).

4.2 STRAIN REACTIVATION AND GROWTH CURVE

The freeze-dried strains of lactic acid bacteria were reactivated in MRS broth (KASVI) and pre-incubated in an oven for 30 minutes to adapt the microorganisms to the growth medium. After reconstitution, about half of the reconstituted volume will be separated for storage. To do this, the separated volume will be transferred to another container containing glycerol (20 % of the total amount), previously sterile and heated to 37°C to reduce viscosity, thus facilitating its dilution and incorporation into the inoculum's MRS broth, which is called a stock solution (SAEKI; FARHAT; PONTES, 2015).The strains used were in capsules with a concentration of 10^9 CFU/g and were added to erlenmeyer flasks containing 300mL of MRS broth, with the initial concentration (T0) for the growth curve reduced to 10^6 CFU/mL.After preparation, the culture was incubated at 35-37°C in anaerobiosis for 48 hours. Points were taken to assess growth at times 0, 2, 4, 8, 12, 18 and 24,36, 48 hours.At each time point, 4mL of the culture was removed t o assess turbidity and 1mL to confirm the viable cell count on plates. After homogenising the working solution, the aliquot of the culture was diluted in 0.1% peptone saline solution (ssp 0.1%) to a dilution of 10^{-8} . The dilutions were inoculated onto MRS agar by pour plate and incubated for 96 hours at 35-37°C in an air circulation oven under anaerobiosis. All counting analyses were carried out in triplicate (APHA, 2015).In order to make it easier to check the concentration of the prepared broths, bacterial concentration curves were constructed using the turbidimetry technique, a method that gives an estimate of bacterial growth in real time through optical density measurements. The reading cuvettes were diluted with distilled water and the absorbance measurements were read on a UV/VIS spectrophotometer (SHIMADZU UV-PROBE 1800) at a wavelength of 600nm. The correlation graph between cell count on plates and turbidity was drawn up for each microorganism (BEGOT et al., 1996; BOVO, 2013).

4.3 PREPARATION OF INOCULUM WITH STANDARDISED CONCENTRATION

The bacterial inoculum of standardised concentration was produced by incubating the microorganisms in 300mL of MRS broth for the time defined by the concentration curve, and the bacterial count in the inoculum was confirmed by correlating it with the absorbance. After the incubation period, the inoculum was fractionated into sterile centrifuge tubes with a final volume of 10mL and processed in different ways based on the analyses carried out.For the lactose metabolisation tests, 1mL of the inoculum was used and added directly to the tubes of lactose broth.For the antimicrobial susceptibility tests, 1mL was taken from the centrifuge tube and diluted in 0.1% ssp (1:9). For the adsorption and biomethabolisation tests, the inoculum was centrifuged at 1800 rpm for 15 minutes to pellet the bacterial cells. After removing the supernatant, the pellet was washed using 0.1% ssp and centrifuged again. The procedure was repeated twice and after the second wash, the final pellet was resuspended in the working buffer solutions.

4.4 ADSORPTION AND BIOMETABOLISATION OF AFLATOXIN B1 AND ZEARALENONE BY SACCHAROMYCES CEREVISAE

S. cerevisiae strains LL74, LL08 and LL83 were reactivated as described in section 4.2. The strains were propagated at 30ºC in centrifuge tubes containing 10 mL of YPD broth for two days. The optical densities of the samples were determined at 600nm and adjusted to 2.0 with sterile distilled water. All the strains were kept active in YPD broth throughout the study period and preserved at 4°C.

The yeasts were first cultivated in 250mL of YPD at 30°C using an orbital shaker (150 rpm). The culture broths were then centrifuged at 10,000 rpm for 10 min, the supernatants were discarded and the pellets frozen at -80°C and freeze-dried to obtain dry cells, as described in section 4.3.

4.4.1 **Evaluations of the adsorptive capacity of Saccharomyces cerevisiae against aflatoxin B1 and zearalenone**

The S. cerevisiae strains were tested in YPD broth supplemented with 1,261 µg/mL of each mycotoxin (MT), YPD+AF and YPD+ZEN. Centrifuge tubes containing 5 mL of YPD+MT were inoculated in triplicate with 0.1 mL of each inoculum. Negative controls were prepared with 0.1 mL of sterile distilled water.

The adsorption of mycotoxins to the yeast cell wall was confirmed using a static gastrointestinal model. The simulation solutions consisted of a physiological solution and enzymes: for the gastric simulation, NaCl 125mmol/L, KCl 7mmol/L, NaHCO3 45mmol/L and pepsin 3g/L (porcine gastric mucosa, 800-2500 U/mg) at pH 3; for the intestinal simulation: bile 0.5% (w/v), trypsin type IX-S 1mg/mL (13000-20000 BAEE U/mg) and α-chymotrypsin type II 1mg/mL (pancreas, ≥ 40 U/mg) at pH 6. The reaction solutions were prepared at the time of use and supplemented with the target concentrations of MT.

The experiments were conducted by adding the lyophilised inoculum to the solutions, making working concentrations equivalent to 0%, 25%, 50%, 75% and 100% (0, 0.5, 1.0, 1.5 and 2.0 mg/mL), followed by incubation for 1 h at 37°C, under agitation (150 rpm). After this period, the solutions were then centrifuged at 10,000 rpm for 10 min, the supernatants collected and the MT concentrations determined by HPLC-FL as described below. The adsorption percentages of MT yeast cells were calculated using the equation below: (KELLER et al., 2015).Adsorption (%) = Concentration of AFM1 in the supernatantX 100 Concentration of AFM1 in the positive control

4.4.2 **Mycotoxin extraction and detection**

The culture media and gastrointestinal simulation solutions were extracted with an equivalent volume of acetonitrile/methanol/acetic acid (78:20:2, v/v/v) by vortex shaking for 1 min. The samples (2mL) were then collected and filtered through 0.45µm syringe-fit PP filters (Merck®) into clean amber borosilicate glass vials. The cell pellets were extracted with 1mL of the same solution mixed with 1mL of ultrapure water under agitation for 1h. After filtering, the samples were preserved at 4°C until analysed by HPLC.The HPLC system consisted of a Varian Prostar 210® pump (Varian-Agilent®, California, USA), a Varian Prostar 410® autosampler (Varian-Agilent, California, USA) and a Jasco FP-920® fluorescence detector. The instrument and chromatographic data were managed by a Varian 850-MIB® data system interface (Varian-Agilent®, California, USA) and a Galaxie® chromatography data system (Varian-Agilent®, California, USA), respectively.

The AFB1 detection process was conducted as described by da Silva et al. (2015), using a Spherisorb ODS2 Column® (150mm × 4.6mm, 5μm) (Waters®, Massachusetts, USA). The best chromatographic conditions comprised a mixture of methanol:acetonitrile:ultrapure water (2:2:6 v/v/v) as the mobile phase under isocratic elution conditions at 1mL/min with an injection volume of 50μL and a column temperature of 50ºC. Fluorescence detection was adjusted to excitation and emission wavelengths of 365nm and 445nm. The retention time for AFB1 was 20 minutes under these conditions. Calibration curves were prepared with AFB1 standards ranging from 0.084 to 20.376mg/g. In relation to ZEA (KELLER et al., 2015), the samples were eluted at 1.0mL/min by

23 min under isocratic conditions using methanol/water/acetic acid (65:35:1, v/v/v). The chromatographic separations were carried out at 30ºC on a C18 YMC-Pack ODS-AQ® (YMC®, Kyoto, Japan) reversed-phase analytical column (250 x 4.6 mm I.D., 5 μm) equipped with a pre-column employing the same stationary stage. An injection volume of 50μL was applied and fluorescence detection was carried out at $\lambda exc = 236nm$, $\lambda em = 418nm$. The retention times recorded for ZEA were approximately 21 min. Calibration curves were prepared with ZEA standards at a concentration between 0.25 and 2.0μg/mL.

4.4.3 **Biotransformation of mycotoxins by Saccharomyces cerevisiae**

The concentration of biometabolised MT was obtained using the equation below:

MTb = MTi - MTads - MTdes

Where MTb is the concentration of biotransformed mycotoxin after the incubation period, MTi is the initial percentage of mycotoxin present in the assay (BOVO et al., 2013), MTads is the concentration of adsorbed mycotoxin and MTdes is the concentration of desorbed mycotoxin.

4.5 ANTIMICROBIAL SUSCEPTIBILITY TESTING

The antimicrobial susceptibility tests were carried out as described by the Clinical and Laboratory Standards Institute (CLSI) with modifications for minimum inhibitory concentration (MIC) tests (CLSI, 2012). The modifications made were based on the subsequent application of the microorganisms of interest selected in the screening tests.

The antimicrobial concentrations chosen were based on the Maximum Residue Limits for antimicrobials and antiparasitic agents in milk described in the PAMvet (BRASIL, 2003a). The MICs of streptomycin, penicillin G, tetracycline and azithromycin were evaluated against the previously selected Lactobacillus strains.

4.5.1 **Preparation of solutions**

All solutions were prepared using sterile Phosphate Buffered Solution (PBS), pH 7.2 as a diluent.

4.5.1.1 Streptomycin

From a starting solution of streptomycin 3200μg/L. Tubes containing the concentrations 1600, 800, 400, 200, 100, 50, 25, 12.5, 6.25μg/L were prepared by serial dilution. This totalled 10 tubes with different working concentrations of streptomycin.

4.5.1.2 Penicillin G

From a starting solution of penicillin G 64μg/L. Tubes containing the concentrations 32, 16, 8, 4, 2, 1, 0.5, 0.25, 0.125μg/L were prepared by serial dilution. This totalled 10 tubes with different working concentrations of penicillin G.

4.5.1.3 Tetracycline

From a starting solution of tetracycline 1600μg/L. Tubes containing concentrations of 800, 400, 200, 100, 50, 25, 12.5, 6.25 and 3.125μg/L were prepared by serial dilution. This totalled 10 tubes with different working concentrations of tetracycline.

4.5.1.4 Azithromycin

From a starting solution of azithromycin 640μg/L. Tubes containing the concentrations 320, 160, 80, 40, 20, 10, 5, 2.5, 1.25μg/L were prepared by serial dilution. This totalled 10 tubes with different working concentrations of azithromycin.

4.5.2 Experimental procedure

The procedure described below was carried out in a similar way for each of the Lactobacillus strains studied.For each antimicrobial (AB) used, 10 tubes containing 2mL of Mueller-Hinton broth (MH) were labelled according to the antimicrobial and the corresponding concentration.

Next, 1mL of the standardised inoculum (10^8 CFU/mL) obtained in accordance with item 4.3 and 1mL of each of the antimicrobial working solutions were added. After preparation, the tubes were incubated at 35°C for 48 hours in anaerobiosis.

Two controls were prepared for this test. The positive control was prepared by adding 2mL of MH broth, 1mL of the bacterial inoculum and 1mL of 0.1% ssp. For the negative control, 2mL of MH broth and 2mL of 0.1% ssp were added, completing the 4mL of solution in each tube. The controls were incubated under the same conditions as the other tubes. After the incubation period, the tubes were removed and sorted by antimicrobial and in ascending order of concentration. The minimum inhibitory capacity of each antimicrobial was assessed by visually comparing the turbidity and presence of a background body between the tubes and the positive control.

The confirmatory test was carried out using plate counting, where 1mL was aliquoted from the tubes with the highest concentration of each antimicrobial that showed cell growth. Each aliquot was then serially diluted to a dilution of 10^{-10} and inoculated into MRS medium. The plates were incubated at 35°C in anaerobiosis for 96 hours.

4.6 LACTOSE METABOLISATION

To determine lactose metabolisation, lactose broth (Difco®, Becton Dickinson, New Jersey, USA) containing 0.5mg of lactose per 100mL of medium was used, fractionated into tubes containing 10mL of broth.

1mL of the standardised microbial inoculum (10^8 CFU/mL) obtained in accordance with item 3.2.2 was added to each tube and incubated for 24 hours at 35°C in anaerobiosis.

4.6.1 **Lactose quantification**

Lactose quantification was carried out as described by Ni, Huang and Kokot (2003) by reducing potassium ferricyanide in an alkaline medium. The pH after the incubation period was also evaluated as a measure of the concentration of lactic acid produced.

After incubation, 5mL of the lactose broth was removed and transferred to a beaker. After checking the initial pH, the sample was neutralised using a few drops of 1.5mol/L NaOH. After neutralisation, the sample was transferred to a 25mL volumetric flask and volumised using distilled water.

2 mL of the sample was removed from the volumetric flask and transferred to 15 mL centrifuge tubes. 3mL of 20% lead acetate was added to precipitate the compounds in solution and a few drops of 10% sodium phosphate to precipitate the lead acetate. After preparation, the sample was volumised to 10mL and taken to the centrifuge at 3000 rpm for 10 minutes. Test tubes were prepared containing 1.4mL of 1.5mol/L NaOH and 2.8mL of distilled water. After homogenising, 0.8mL of 0.4g/100mL potassium ferricyanide was added and shaken vigorously. 0.4mL of the sample was added to the prepared tube and homogenised. The blank solution was prepared as described above, except for the addition of the sample.

The sample and blank solution were then heated at 80°C for 5 minutes and taken while still hot to be read in a spectrophotometer using a wavelength of 420nm. The samples were heated individually to ensure temperature during reading. All samples were made in triplicate.

The calibration curve for lactose was carried out using lactose solutions with concentrations of 0, 0.25, 0.50, 0.75, 1.0, 1.25, 1.50, 1.75, 2.00 mg/mL. The tubes were prepared as described above and the standardised solutions added in place of the sample. The tubes were then heated to 80°C and read in a spectrophotometer at 420nm as described above. The graph to obtain the calibration curve and straight line equation was plotted as abs x g/L lactose.

4.7 ADSORPTION AND BIOMETHABOLISATION OF AFLATOXIN M1

4.7.1 **Sample preparation**

The aflatoxin M1 adsorption and biomethabolisation test was carried out as described by Keller et al. (2015) and Moebus et al. (2023) with adaptations.
The adsorption and biomethabolisation capacities of each microorganism were assessed in two different forms: viable cells with their metabolic apparatus intact and in an inactivated form.
The standard inoculum containing viable cells was obtained as described in section 4.3 and transferred to 1.5 mL centrifuge microtubes. The microtubes were centrifuged and the supernatant removed, leaving only the cell mass in the tube.
The inoculum of inactivated cells was heated at 100°C for 1 hour for complete cell inactivation. This was followed by a similar separation process for the viable cell inoculum.
The aflatoxin M1 working solution was prepared from the concentrated solution with a concentration of 0.5μg/mL (500ppb) solubilised in acetonitrile. Initially, an intermediate solution containing 0.04μg/mL (40ppb) was prepared. Five reaction media were used, varying the composition of the medium and pH in order to assess the adsorptive and biometabolisation effects under different pH conditions. Among the buffer solutions used were phosphate buffers (PBS) at pH 2 and 7 and citrate buffers (CBS) at pH 3 and 6, duly sterilised.
After preparing the solutions, 925μL (9.25mL) of each buffer was added to microtubes containing the pelletised microorganisms. Next, 75μL (0.75mL) of the 0.04μg/mL AFM1 solution was added, giving a final concentration of 0.002μg/mL (2ppb).
The centrifuge microtubes containing the samples were separated and incubated for 4 different times (8, 12, 18 and 24 hours) for each state of cell viability. The entire test was carried out in triplicate.
After each period, the tubes were centrifuged at 1800 rpm for 15 minutes and the supernatant was transferred to a glass vial and taken for quantification in HPLC.
For the desorption test, the same volume of each original buffer was added and resuspended to release the AFM1 molecules bound to the cell structures. After this procedure, the tube was centrifuged again and the supernatant collected in another vial for HPLC quantification.

4.7.2 **Quantification of Aflatoxin M1**

The HPLC system used consists of a Shimadzu LC-20AD pump (Shimadzu®, Kyoto, Jp), a Shimadzu SIL-10AF autosampler (Shimadzu®, Kyoto, Jp), a Shimadzu CTO-20A oven (Shimadzu®, Kyoto, Jp) and a Shimadzu RF-20A fluorescence detector (Shimadzu®, Kyoto, Jp). The samples were separated using a Spherisorb ODS2 Column® (150 mm × 4.6 mm, 5 µm) (Waters®, Massachusetts, USA). The quantification of AFM1 was carried out according to the method proposed by Shundo and co-workers (2004). The chromatographic condition used was: aqueous solution of 2% acetic acid:acetonitrile:methanol (40:35:25, v/v) in a low pressure system with an injection flow of 1.0mL/min, injection volume of 50µL and oven temperature of 40°C. Fluorescence detection was carried out using an excitation wavelength of 365nm and an emission of
445. The retention time for AFM1 was 2 minutes under these conditions. The calibration curve where the volume of each injection of 50 µL corresponded to 0.01; 0.05; 0.5; 1.0; 2.0; 4.0 ng. The limit of detection (LOD) was 0.01 ng/mL and the limit of quantification (LOQ) was 0.05 ng/mL.

4.7.3 **Calculation of adsorption and biomethabolisation of AFM1**

The adsorptive capacity of each strain in a given buffer was defined using the equation below (KELLER et al., 2015; MOEBUS et al., 2023):Adsorption (%) = Concentration of AFM1 in the supernatantX 100 Concentration of AFM1 in the positive control
The concentration of biometabolised AFM1 was obtained using the equation below:
MTb = MTi - MTads - MTdes
Where MTb is the concentration of biotransformed mycotoxin after the incubation period, MTi is the initial percentage of mycotoxin present in the test, MTads is the concentration of adsorbed mycotoxin and MTdes is the concentration of desorbed mycotoxin.

4.8 ANTIMICROBIAL ADSORPTION AND BIOMETABOLISATION TEST

4.8.1 **Sample preparation**

The antimicrobial adsorption and biomethabolisation test was carried out as described by Keller et al. (2015), Massoud et al. (2020) and Moebus et al. (2023).The adsorption and biomethabolisation capacities of each microorganism were assessed in two different forms: viable cells with their metabolic apparatus intact and in an inactivated form.The standard inoculum containing viable cells was obtained as described in item 4.3 and transferred to 15 mL centrifuge tubes. The tubes were centrifuged and the supernatant removed, leaving only the cell mass in the tube.The inoculum of inactivated cells was heated at 100°C for 1 hour for complete cell inactivation. This was followed by a similar separation process for the viable cell inoculum.Five reaction media were used, varying the composition of the medium and pH in order to assess the adsorptive and biometabolisation effects under different pH conditions. Among the buffer solutions used were phosphate buffers (PBS) at pH 2 and 7 and citrate buffers (CBS) at pH 3 and 6, duly sterilised.Each antimicrobial working solution was prepared as described in 3.2.4.1, changing only the buffers used. The working concentrations were chosen based on those described in PAMvet (BRASIL, 2003a), as follows: Streptomycin, 0.05; 0.1; 0.2; 0.4

μg/mL. Penicillin G, 0.001; 0.002; 0.004; 0.008 μg/mL. Tetracycline, 0.025; 0.05; 0.1; 0.2 μg/mL and Azithromycin, 0.01; 0.02; 0.04; 0.08 μg/mL. 10mL of each solution was added to each centrifuge tube containing the bacterial inoculums.

The centrifuge tubes containing the samples were separated and incubated for 24 hours for each state of cell viability. The entire test was carried out in triplicate.

After each period, the tubes were centrifuged at 1800 rpm for 15 minutes and the supernatant was transferred to a microtube with a screw cap and processed according to the methodologies below.

For the desorption test, the same volume of each original buffer was added and resuspended to release the antimicrobial molecules bound to the cell structures. After this procedure, the tube was centrifuged again and the supernatant collected in another microtube with a screw cap and processed according to the methodologies below.

4.8.2 **Quantification of streptomycin**

Quantification of streptomycin in the samples was carried out as described in the Brazilian Pharmacopoeia for dosing the drug (BRASIL, 2019).

After the incubation period, 5 mL of each sample was transferred to 25 mL volumetric flasks. 1 mL of 1M sodium hydroxide was added to each flask and heated in a water bath for 4 minutes, then cooled rapidly to room temperature.

In each flask, 2mL of a 2% (w/v) ammoniacal ferric sulphate solution diluted in 0.5 M sulphuric acid was added. The flasks were then volumised, homogenised and left to stand for 10 minutes. The samples were then taken to the UV/VIS-PROBE 1800 spectrophotometer (Shimadzu®, Kyoto, Jp) and their absorbances read at a wavelength of 520nm, using the blank solution to adjust the zero. All readings were taken in triplicate.

The calibration curve was drawn up considering the working concentrations and two extrapolated points: 0.01; 0.05; 0.1; 0.2; 0.4; 0.8 μg/mL.

4.8.3 **Quantification of penicillin G**

The quantification of penicillin G in the samples was carried out as described in the Brazilian Pharmacopoeia for dosing the drug (BRASIL, 2019).

The HPLC system used consists of a Shimadzu LC-20AD pump (Shimadzu®, Kyoto, Jp), a Shimadzu SIL-10AF autosampler (Shimadzu®, Kyoto, Jp), a Shimadzu CTO-20A oven (Shimadzu®, Kyoto, Jp) and a DAD SPD-M20A ultraviolet detector (Shimadzu®, Kyoto, Jp).

The samples were separated using a Spherisorb ODS2 Column® (150 mm × 4.6 mm, 5 μm) (Waters®, Massachusetts, USA).
The chromatographic condition used was: aqueous solution of water:acetonitrile:glacial acetic acid (650:350:5.75, v/v) in a low pressure system with an injection flow of 1.0mL/min, injection volume of 50μL. Ultraviolet detection was carried out using a wavelength of 254nm. The retention time for penicillin G was 0.8 minutes under these conditions. The calibration curve where the volume of each injection of 50 μL corresponded to 0.01; 0.05; 0.5; 1.0; 2.0; 4.0 ng. The calibration curve was drawn up considering the working concentrations and two extrapolated points: 0.0005; 0.001; 0.002; 0.004; 0.008; 0.016 μg/mL. All readings were taken in triplicate.

4.8.4 **Quantification of tetracycline**

The quantification of tetracycline in the samples was carried out as described in the Brazilian Pharmacopoeia for dosing the drug (BRASIL, 2019).
After the incubation period, 3 mL of each sample was transferred to 100 mL volumetric flasks. Each flask was filled with 0.25M sodium hydroxide, homogenised and left to stand for six minutes. The samples were then taken to the UV/VIS-PROBE 1800 spectrophotometer (Shimadzu®, Kyoto, Jp) and their absorbances read at a wavelength of 380nm, using the blank solution to adjust the zero. All readings were taken in triplicate.
The calibration curve was drawn up considering the working concentrations and two extrapolated points: 0.0125; 0.025; 0.05; 0.1; 0.2; 0.4 μg/mL

4.8.5 **Quantification of azithromycin**

The quantification of tetracycline in the samples was carried out as described in the Brazilian Pharmacopoeia for dosing the drug (de PAULA et al., 2019).
Aliquot 500μL (0.5mL) into a 50mL volumetric flask and add 5.0mL of the 1.2mg/mL alizarin solution. Make up to the mark with distilled water. The samples were then taken to the UV/VIS-PROBE 1800 spectrophotometer (Shimadzu®, Kyoto, Jp) and their absorbances read at a wavelength of 525nm, using the blank solution to adjust the zero. All readings were taken in triplicate.The calibration curve was drawn up considering the working concentrations and two extrapolated points: 0.005; 0.01; 0.02; 0.04; 0.08; 0.16μg/mL

4.8.6 **Calculation of adsorption and biometabolisation of antimicrobials**

The adsorptive capacity of each strain in a given buffer was defined, similarly to the calculation for AFM1, using the equation below (KELLER et al., 2015; MOEBUS et al., 2023):Adsorption(%) = Concentration ofAB in the supernatantX 100 Concentration of AB in the positive control

The concentration of biometabolised AFM1 was obtained using the equation below:

ABb = ABi - ABads - ABdes

Where ABb is the concentration of biotransformed mycotoxin after the incubation period, ABi is the initial percentage of mycotoxin present in the test, ABads is the concentration of adsorbed mycotoxin and ABdes is the concentration of desorbed mycotoxin.

4.9 STATISTICAL ANALYSES

Results are expressed as mean and mean standard error with triplicate determinations. Statistical analyses were carried out using InfoStat software version 2020 (InfoStat Software, Córdoba, Argentina). In vitro gastrointestinal transit tolerance results and microbial growth kinetics were analysed using Spearman's correlation analysis. Bacterial adhesion capacities, lactose consumption, mycotoxins and antimicrobial bioremotion were analysed using one-way analysis of variance (ANOVA) with Tukey's test as a post-hoc test. Differences were considered statistically significant at $p \geq 0.05$.

5 DEVELOPMENT

5.1 IN VITRO MYCOTOXIN DECONTAMINATION BY SACCHAROMYCES CEREVISIAE STRAINS ISOLATED FROM BOVINE FORAGE

Submitted to: Fermentation special issue: Health and Bioactive Compounds of Fermented Foods and By-ProductsSubmission date: 07 June 2023 Acceptance date: 19 June 2023 Publication date: 22 June 2023 The following text complies with the journal's formatting standards as described in the Instructions for Authors section, available at: https://www.mdpi.com/journal/fermentation/instructions

In Vitro Mycotoxin Decontamination by Saccharomyces cerevisiae Strains Isolated from Bovine Forage Victor Farias Moebus 1,*,†, Leonardo de Assunção Pinto 2,†, Felipe Braz Nielsen Köptcke 3,†, Kelly **Moura Keller [4,†], Marcos Aronovich [5,†] and Luiz Antonio Moura Keller [6,†]**

[1] Veterinary Hygiene and Technology of Products of Animal Origin, Universidade Federal Fluminense,
Niterói 24230-321, Rio de Janeiro, Brazil

[2] Plant Biotechnology and Bioprocesses, Federal University of Rio de Janeiro, Rio de Janeiro 21941-590, Rio de Janeiro, Brazil; leodeasspinto@gmail.com

[3] Pharmacy Course, Universidade Federal Fluminense, Niterói 24241-000, Rio de Janeiro, Brazil; felipekoptcke@id.uff.br

[4] Department of Preventive Veterinary Medicine, Veterinary School, University of Lisbon Federal de Minas Gerais, Belo Horizonte 31270-901, Minas Gerais, Brazil; kelly.medvet@gmail.com

[5] Agricultural Research Company of the State of Rio de Janeiro (PESAGRO- RIO), Niterói 24120-191, Rio de Janeiro, Brazil; marcosaronovich@gmail.com

[6] Veterinary Medicine Course, Universidade Federal Fluminense, Niterói 24230-321, Rio de Janeiro, Brazil; luiz_keller@id.uff.br

† These authors contributed equally to this work.

* Correspondence: victormoebus@id.uff.br; Tel: +55-21985822408

Abstract: Aflatoxin B1 (AFB1) and Zearalenone (ZEN) are among the most common and important mycotoxin contaminants in agricultural products, with AFB1 comprising a liver carcinogen and ZEN responsible for reproductive dysfunctions. As mycotoxins are heat-stable, their removal is carried out mainly by anti-mycotoxin additives. This includes the yeast Saccharomyces cerevisiae. In this context, this study aimed to evaluate the in vitro detoxification of AFB1 and ZEN at pH 3 and 6 by three S. cerevisiae strains isolated from bovine forage, coded LL74, LL08, and LL83, determining the adsorption and biotransformation capacities of each strain. The yeast were freeze-dried and added, in triplicate, at 0, 0.5, 1.0, 1.5, and 2.0

mg mL^{-1} to a static gastrointestinal model. Final mycotoxin concentrations were

determined by HPLC-FL. The evaluated strains exhibited high mycotoxin adsorption rates

(20-55%), especially the LL08 strain, although low biotransformation, both equivalent to a commercial strain. The results indicate that pH does not interfere in
AFB1 detoxification, while the use of 2.0 mg mL^{-1} of the LL83 S. cerevisiae strain led to higher ZEN adsorption at pH 3. The investigated strains indicate the possibility for use in in vivo conditions and high potential for commercial applications, with LL08 as the most promising strain.

Keywords: aflatoxin; zearalenone; anti-mycotoxins additive; adsorption; biotransformation

1. Introduction

Mycotoxins (MT) are produced by the secondary metabolism of fungi and are toxic to both humans and animals through inhalation, contact, and ingestion. Mycotoxin contamination is a recurring foodstuff problem, occurring in the field and/or during the harvest, storage, processing, or feeding stages [1], with some of the most common mycotoxins in agricultural products comprising aflatoxins (AF), zearalenone (ZEN), deoxynivalenol (DON), fumonisins (FM), and ochratoxin A (OTA) [2].
Aflatoxins are highly toxic, stable, and resistant molecules, being detectable at various levels of the production chain, representing one of the major food safety concerns worldwide. Aflatoxins alone represented 20.9% of food and feed notifications in Europe since 2002. The four most noteworthy AF are B1, B2, G1, and G2, all of which exhibit toxic effects [3,4].
Despite being reported worldwide, the occurrence of AF in food, feed and milk is a major problem in tropical and subtropical regions, mainly in developing countries, where climatic conditions favour fungal growth and AF production. Requiring the implementation of food quality control programmes to avoid this risk [5-7].
The AFB1 is considered the most potent of then, being responsible to several physiological damages to purification and metabolisation organs. The AFB1 can also be associated with the occurring of liver neoplasis, being considered the most potent natural liver carcinogen for some species [8-10].
Lactating animals continuously exposed to AFB1 ingestion may also produce a metabolite excreted in milk, aflatoxin M1 (AFM1). The AMF1 is associated with liver cancer and other acute pathologies when ingested by humans [5,8,9,11]. Zearalenone is a heat-stable, nonsteroidal estrogenic compound produced by many Fusarium species, with F. graminearum as the most noteworthy in this regard [12,13]. Food ZEN contamination has been reported worldwide, especially in temperate climates. Zearalenone concentrations are usually quite low in grains in the field, increasing under 30-40% moisture level storage conditions. Biologically, ZEN and its derivatives exhibit estrogenic activity, leading to reproductive dysfunctions in intoxicated animals [12,13].
Numerous strategies have been proposed to counteract toxic mycotoxin effects, including the use of anti-mycotoxin additives (AMA) such as inorganic and organic binding agents. Anti-mycotoxin additives comprise a group of products able to neutralise, adsorb and inactivate mycotoxins in the gastrointestinal tract of animals, and their effects are determined by their components and mechanism of action [14].
Many AMA focusing on mycotoxin removal have been developed, most based on inorganic binders capable of removing a high number of mycotoxins. The use of microorganisms

displaying performance improvement properties, however, especially Saccharomyces cerevisiae, has increased in recent years due to beneficial effects associated with their ingestion, leading to increased dry matter intake and milk production and reduced risks for ruminal acidosis, as well as probiotic effects [15,16].

Numerous studies have demonstrated the AMA capacity of S. cerevisiae, due to its cell wall composition, resulting in mycotoxin adsorption. The cell wall compositions of this yeast play a major role in mycotoxin binding, exhibiting a strain-dependent performance also sensitive to in vivo conditions, such as animal gastrointestinal tract pH variations [17-20]. These components are essential for the selection of strains used in commercial AMA, optimised for potent adsorbent effects without altering beneficial effects.

The genotypic variation of each S. cerevisiae strain can, however, alter these components, affecting mycotoxin adsorption efficiency. Studies focused on the genetic improvement of this yeast and the discovery of wild strains with high adsorptive potential have increased to support the development of new commercial AMA and improve already marketed AMA [21,22]. Industries can apply these strains to raw milk that would otherwise be discarded due to mycotoxin contamination exceeding established limits. The same strains may also optimise the fermentation process required in dairy product production, inserting probiotics both beneficial and attractive to consumers in the final product [23].

In this context, the present study aimed to determine the adsorption and biodegradation capacities of three S. cerevisiae strains isolated from bovine forage with regard to AFB1 and ZEN detoxification compared to a commercial strain.

2. Materials and Methods

2.1. Biological Material

Three S. cerevisiae strains isolated from bovine forage, coded S. cerevisiae LL74, LL08, and LL83, were investigated. The strains were deposited at the Mycology and Mycotoxin Laboratory culture collection belonging to the Universidade Federal de Minas Gerais (LAMICO-UFMG) and the National University of Río Cuarto (Córdoba, Argentina) collection centre. A commercial S. cerevisiae strain (coded CS) was used as the control, applied at the recommended dose of 2.0 mg mL^{-1} (2 mg g^{-1} per feed, with a cell count ≥ 10^7 CFU g^{-1}). The S. cerevisiae strains were propagated at 30 °C for two days in centrifuge tubes containing 10 mL of YPD broth (1% yeast extract, 2% peptone, and 2% glucose). The optical densities of the samples were determined at 600 nm and adjusted to 2.0 with sterile distilled water. All strains were maintained active in the YPD broth throughout the study period and preserved at 4 °C.

2.2. Mycotoxin Saccharomyces Cerevisiae Adsorption and Desorption Assessments

The Saccharomyces cerevisiae strains were tested in YPD broth supplemented with 1261 μg mL^{-1} of each mycotoxin (MT) (Sigma-Aldrich®, Missouri, USA), as YPD + AF and YPD + ZEN. Centrifuge tubes containing 5 mL of YPD + MT were inoculated in triplicate in 0.1 mL of the inoculum. Negative controls were prepared using 0.1 mL of sterile distilled water.The yeast were first cultivated in 250 mL YPD at 30 °C for 48 h as requested in point 3 using an orbital shaker (150 rpm). The culture broths were then centrifuged at

10,000 rpm for 10 min, the supernatants were discarded and the pellets frozen at -80 ° C and freeze-dried, obtaining dried cells.Mycotoxin yeast cell wall adsorption was confirmed using a static gastrointestinal model. The simulation solutions were composed of a physiological solution and enzymes, namely 125 mmol L^{-1} NaCl, 7 mmol L^{-1} KCl, 45 mmol L^{-1} NaHCO3 and 3 g L^{-1} pepsin (porcine gastric mucuos, 800-2500 U mg^{-1}) at pH 3 for the gastric simulation and 0.5% bile (w/v), 1 mg mL^{-1} trypsin type IX-S (13,000-20,000 BAEE U mg^{-1}), and 1 mg mL^{-1} α-chymotrypsin type II (pancreas, ≥40 U mg^{-1}) at pH 6 for the intestinal simulation. The reaction solutions were prepared at the time of use and supplemented with 1261 µg mL^{-1} of each MT evaluated. Considering the recommended dose for using commercial products that have a yeast concentration ≥ 10^7 CFU g^{-1} (2.0 mg mL^{-1}), the inoculum concentrations used in the assay corresponded to 0%, 25%, 50%, 75% and 100% of the recommended dose (0, 0.5, 1.0, 1.5 and 2.0 mg mL^{-1}). The freeze-dried yeast powder was added to reaction tubes and incubated at 37 °C for 1 h (150 rpm). After incubation period, the solutions were then centrifuged at 10,000 rpm for 10 min, the supernatants collected and the MT concentrations determined by HPLC-FL as described in Section 2.3. The MT yeast cell adsorption percentages were calculated using Equation (1) [15]. MT concentration in *the* supernatant MT concentration in *the* positive control The desorption assay, was proceeded washed the yeast pellet with YPD and resuspended. The solutions were then centrifuged at 10,000 rpm for 10 min, the supernatants collected and the desorbed MT concentrations was determined by HPLC-FL as described in Section 2.3. The MT yeast cell desorption percentages were calculated using Equation (1) [15]. A variance analysis (ANOVA) was performed on the data by applying Fisher's LSD test and used to compare the adsorption means of the two pH conditions employing the PROC GLM SAS package (SAS Institute, Cary, NC, USA).

2.3. Mycotoxin Extraction and Detection

The HPLC system comprised a Varian Prostar 210® pump (Varian-Agilent®, Palo Alto, CA, USA), a Varian Prostar 410® autosampler (Varian-Agilent, Palo Alto, CA, USA), and a Jasco FP-920® fluorescence detector. The instrument and chromatographic data were managed by a Varian 850-MIB® data system interface (Varian-Agilent®, Palo Alto, CA, USA) and a Galaxie® chromatography data system (Varian-Agilent®, Palo Alto, CA, USA), respectively. The AFB1 detection process was carried out using a Spherisorb ODS2 Column® (150 mm × 4.6 mm, 5 µm) (Waters®, Milford, MA, USA, USA) [24]. The best chromatographic conditions comprised a mixture of methanol:acetonitrile:ultrapure water (2:2:6 v/v/v) as the mobile phase under isocratic elution conditions at 1 mL min^{-1} with a 50 µL injection volume and 50 °C column oven temperature. Fluorescence detection was set at 365 nm and 445 nm excitation and emission wavelengths. The AFB1 retention time was 20 min under these conditions. Calibration curves were prepared with AFB1 standards ranging from 0.084 to 20.376 mg g^{-1} (Sigma-Aldrich®, St. Louis, MO, USA, USA). According to ZEN [15], samples were eluted at 1.0 mL min^{-1} for 23 min under isocratic conditions using methanol:water:acetic acid (65:35:1, v/v/v).

Chromatographic separations were performed at 30 °C on a C18 reversed-phase YMC-Pack

ODS-AQ® (YMC®, Kyoto, Japan) analytical column (250X4.6 mm I.D., 5 μm) fitted with a pre-column employing the same stationary phase. A 50 μL injection volume was applied and the fluorescence detection was performed at λexc = 236 nm, λem = 418 nm, and gain = 1000. Recorded retention times for ZEN were approximately 21 min. Calibration curves were prepared with ZEN standards (Sigma-Aldrich®, St. Louis, MO, USA, USA) at a concentration between 0.25 and 2.0 μg mL-1. Saccharomyces Cerevisiae Mycotoxin Biotransformation Biotransformed MT concentrations were determined by the difference between adsorption and desorption assay values, according to Equation (2):

MTb = MTi - MTads - MTdes

where MTb is the biotransformed mycotoxin concentration following the incubation period, MTi is the initial mycotoxin percentage present in the assay [25], MTads is the adsorbed mycotoxin concentration and MTdes is the desorbed mycotoxin concentration. A variance analysis (ANOVA) was performed by applying Fisher's LSD test and used to compare the means of biotransformed mycotoxins at the two investigated pH values using the PROC GLM SAS package (SAS Institute, Cary, NC, USA).

3. Results and Discussion

The yeast results at different concentrations and conditions obtained in the present study are discussed as both adsorptive potential and biotransformation potential. All investigated strains exhibited high mycotoxin adsorption rates but a lower biotransformation capability concerning AF and ZEN.

3.1. Adsorptive Potential

The adsorption capacities of each strain following the incubation period are depicted in Figures 1-6.

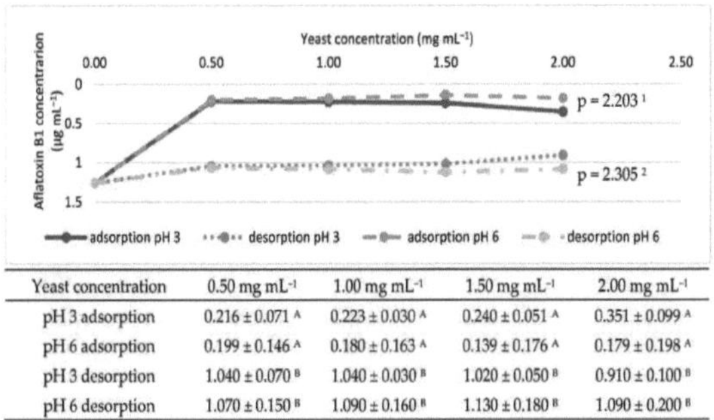

Yeast concentration	0.50 mg mL-1	1.00 mg mL-1	1.50 mg mL-1	2.00 mg mL-1
pH 3 adsorption	0.216 ± 0.071 A	0.223 ± 0.030 A	0.240 ± 0.051 A	0.351 ± 0.099 A
pH 6 adsorption	0.199 ± 0.146 A	0.180 ± 0.163 A	0.139 ± 0.176 A	0.179 ± 0.198 A
pH 3 desorption	1.040 ± 0.070 B	1.040 ± 0.030 B	1.020 ± 0.050 B	0.910 ± 0.100 B
pH 6 desorption	1.070 ± 0.150 B	1.090 ± 0.160 B	1.130 ± 0.180 B	1.090 ± 0.200 B

Figure 1 Adsorption levels and statistical comparisons between AFB1 adsorption and desorption rates by the LL83 Saccharomyces cerevisiae strain under in different pH conditions. [1] Adsorption p value according to the LSD Fisher test comparison at pH 3 and 6; [2] Desorption p value according to the LSD Fisher test comparison at pH 3 and 6; A and B represent differences according to the LSD Fisher test comparison. AFB1 concentrations are expressed in μg mL-1.

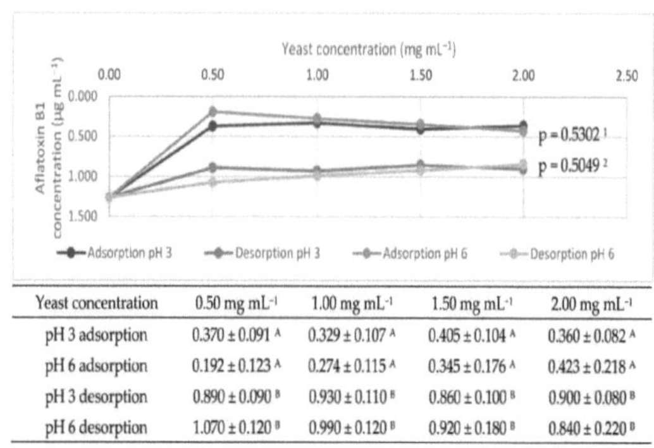

Yeast concentration	0.50 mg mL^{-1}	1.00 mg mL^{-1}	1.50 mg mL^{-1}	2.00 mg mL^{-1}
pH 3 adsorption	0.370 ± 0.091 A	0.329 ± 0.107 A	0.405 ± 0.104 A	0.360 ± 0.082 A
pH 6 adsorption	0.192 ± 0.123 A	0.274 ± 0.115 A	0.345 ± 0.176 A	0.423 ± 0.218 A
pH 3 desorption	0.890 ± 0.090 B	0.930 ± 0.110 B	0.860 ± 0.100 B	0.900 ± 0.080 B
pH 6 desorption	1.070 ± 0.120 B	0.990 ± 0.120 B	0.920 ± 0.180 B	0.840 ± 0.220 B

Figure 2 Adsorption levels and statistical comparison between AFB1 adsorption and desorption rates by the LL08 Saccharomyces cerevisiae strain under different pH conditions. [1] Adsorption p value according to the LSD Fisher test comparison at pH 3 and 6; [2] Desorption p value according to the LSD Fisher test comparison at pH 3 and 6; A and B represent differences according to the LSD Fisher test comparison. AFB1 concentrations are expressed in µg mL-1.

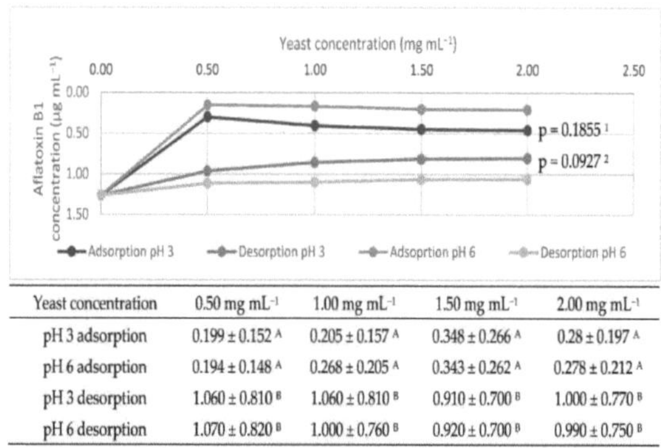

Yeast concentration	0.50 mg mL^{-1}	1.00 mg mL^{-1}	1.50 mg mL^{-1}	2.00 mg mL^{-1}
pH 3 adsorption	0.199 ± 0.152 A	0.205 ± 0.157 A	0.348 ± 0.266 A	0.28 ± 0.197 A
pH 6 adsorption	0.194 ± 0.148 A	0.268 ± 0.205 A	0.343 ± 0.262 A	0.278 ± 0.212 A
pH 3 desorption	1.060 ± 0.810 B	1.060 ± 0.810 B	0.910 ± 0.700 B	1.000 ± 0.770 B
pH 6 desorption	1.070 ± 0.820 B	1.000 ± 0.760 B	0.920 ± 0.700 B	0.990 ± 0.750 B

Figure 3. Adsorption levels and statistical comparison between AFB1 adsorption and desorption rates by the LL74 Saccharomyces cerevisiae strain under different pH conditions. [1] Adsorption p value according to the LSD Fisher test comparison at pH 3 and 6; [2] Desorption p value according to the LSD Fisher test comparison at pH 3 and 6; A and B represent differences according to the LSD Fisher test comparison. AFB1 concentrations are expressed in µg mL-1.

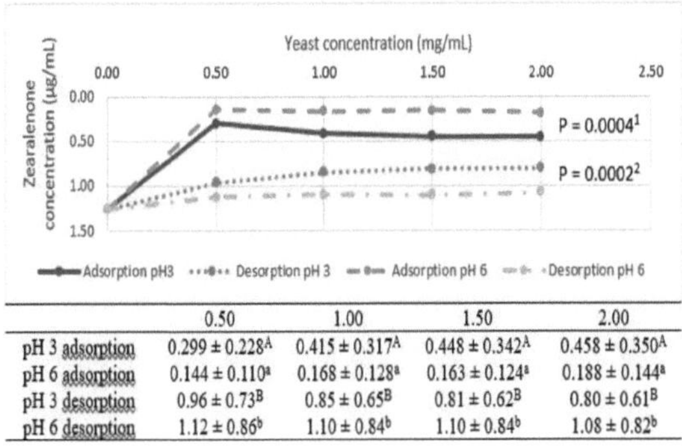

	0.50	1.00	1.50	2.00
pH 3 adsorption	0.299 ± 0.228^{A}	0.415 ± 0.317^{A}	0.448 ± 0.342^{A}	0.458 ± 0.350^{A}
pH 6 adsorption	0.144 ± 0.110^{a}	0.168 ± 0.128^{a}	0.163 ± 0.124^{a}	0.188 ± 0.144^{a}
pH 3 desorption	0.96 ± 0.73^{B}	0.85 ± 0.65^{B}	0.81 ± 0.62^{B}	0.80 ± 0.61^{B}
pH 6 desorption	1.12 ± 0.86^{b}	1.10 ± 0.84^{b}	1.10 ± 0.84^{b}	1.08 ± 0.82^{b}

Figure 4. Adsorption levels and statistical comparison between ZEN adsorption and desorption rates by the LL83 Saccharomyces cerevisiae strain under different pH conditions. [1] Adsorption p value according to the LSD Fisher test comparison at pH 3 and 6; [2] Desorption p value according to the LSD Fisher test comparison at pH 3 and 6; A, a, B and b represent differences according to the LSD Fisher test comparison. ZEN concentrations are expressed in µg mL-1.

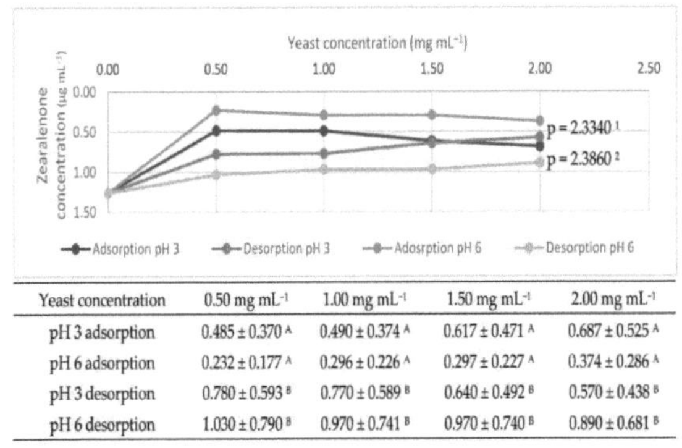

Yeast concentration	0.50 mg mL^{-1}	1.00 mg mL^{-1}	1.50 mg mL^{-1}	2.00 mg mL^{-1}
pH 3 adsorption	0.485 ± 0.370^{A}	0.490 ± 0.374^{A}	0.617 ± 0.471^{A}	0.687 ± 0.525^{A}
pH 6 adsorption	0.232 ± 0.177^{A}	0.296 ± 0.226^{A}	0.297 ± 0.227^{A}	0.374 ± 0.286^{A}
pH 3 desorption	0.780 ± 0.593^{B}	0.770 ± 0.589^{B}	0.640 ± 0.492^{B}	0.570 ± 0.438^{B}
pH 6 desorption	1.030 ± 0.790^{B}	0.970 ± 0.741^{B}	0.970 ± 0.740^{B}	0.890 ± 0.681^{B}

Figure 5. Adsorption levels and statistical comparison between ZEN adsorption and desorption rates by the LL08 Saccharomyces cerevisiae strain under different pH conditions. [1] Adsorption p value according to the LSD Fisher test comparison at pH 3 and 6; [2] Desorption p value according to the LSD Fisher test comparison at pH 3 and 6; A and B represent differences according to the LSD Fisher test comparison. ZEN concentrations are expressed in µg mL-1.

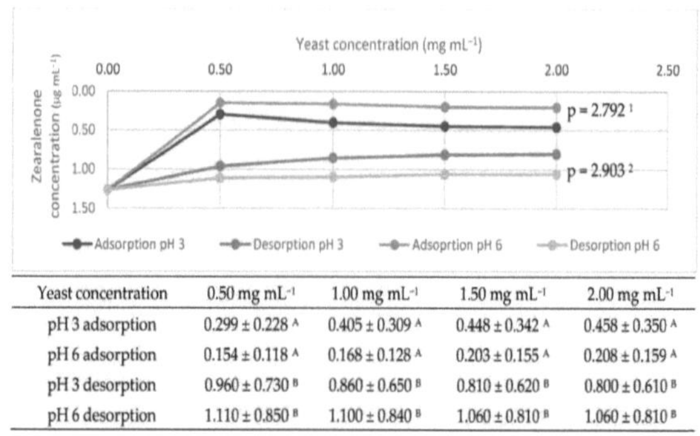

Yeast concentration	0.50 mg mL-1	1.00 mg mL-1	1.50 mg mL-1	2.00 mg mL-1
pH 3 adsorption	0.299 ± 0.228 A	0.405 ± 0.309 A	0.448 ± 0.342 A	0.458 ± 0.350 A
pH 6 adsorption	0.154 ± 0.118 A	0.168 ± 0.128 A	0.203 ± 0.155 A	0.208 ± 0.159 A
pH 3 desorption	0.960 ± 0.730 B	0.860 ± 0.650 B	0.810 ± 0.620 B	0.800 ± 0.610 B
pH 6 desorption	1.110 ± 0.850 B	1.100 ± 0.840 B	1.060 ± 0.810 B	1.060 ± 0.810 B

Figure 6. Adsorption levels and statistical comparison between ZEN adsorption and desorption rates by the LL74 Saccharomyces cerevisiae strain under different pH conditions. [1] Adsorption p value according to the LSD Fisher test comparison at pH 3 and 6; [2] Desorption p value according to the LSD Fisher test comparison at pH 3 and 6; A and B represent differences according to the LSD Fisher test comparison. ZEN concentrations are expressed in µg mL-1.

Gradual increases in mycotoxin adsorption rates with increasing yeast cell counts were noted, albeit with no significant difference for AFB1 ($p > 0.05$). With regard to ZEN, only the LL83 strain applied at 2.0 mg mL-1 was significantly different when comparing both pH values ($p < 0.05$). Mycotoxin adsorption occurs through hydrophobic and electrostatic interactions between the metabolite and the binding sites formed by the yeast cell constituents, such as weak hydrogen bonds and van der Waals bonds. Thus, more cells lead to more binding sites and, consequently, higher MT adsorption rates [23,26]. The differences observed between the evaluated yeast strains is due to innate variations among yeast species, due to a highly diversified cell wall composition. Consequently, the adsorption capacities of the different strains may vary under the same conditions [23]. Adsorption rates can also be affected by yeast cell wall thickness, as yeast with thick cell walls present more surface binding sites when cultivated in certain growth media [21,26]. Besides adsorptive effects associated with cell wall components, other cellular mechanisms related to mycotoxin removal are also noteworthy, such as MT biotransformation catalysed by complex chemical reactions. This reinforces that models that jointly consider all these factors are required for adequate AMA product development protocol and optimisation [27]. Figure 7 compares the commercial and isolated S. cerevisiae strains applied at 2.0 mg mL-1 for each pH condition evaluated herein.

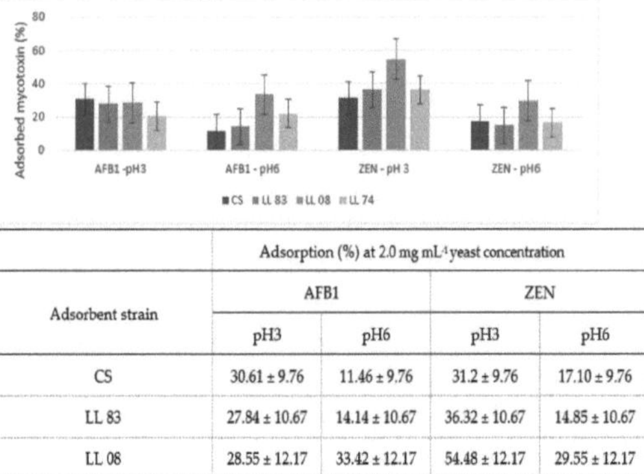

Adsorbent strain	Adsorption (%) at 2.0 mg mL^{-1} yeast concentration			
	AFB1		ZEN	
	pH3	pH6	pH3	pH6
CS	30.61 ± 9.76	11.46 ± 9.76	31.2 ± 9.76	17.10 ± 9.76
LL 83	27.84 ± 10.67	14.14 ± 10.67	36.32 ± 10.67	14.85 ± 10.67
LL 08	28.55 ± 12.17	33.42 ± 12.17	54.48 ± 12.17	29.55 ± 12.17
LL 74	20.46 ± 8.72	21.96 ± 8.72	36.32 ± 8.72	16.43 ± 8.72

Figure 7. Mycotoxin adsorption of commercial and isolated S. cerevisiae strains at 2.0 mg mL-1 under gastric and intestinal simulation conditions at pH 3 and 6. p value > 0.05.

Better results are generally expected for commercial strains, as they are optimised to adsorb mycotoxins in both acidic and neutral conditions and can act throughout the entire gastrointestinal tract. However, the isolated strains evaluated herein exhibited similar adsorption rates at both pH values when compared to the applied CS, albeit with no significant difference ($p > 0.05$).The LL 08 strain obtained better results than the CS in all evaluated conditions, with only a slightly lower AFB1 adsorption at pH 3, albeit with no significant difference ($p > 0.05$). The other obtained adsorption values were higher than the CS, especially ZEN adsorption at pH 3, which was 57% higher than CS, (p value = 0.0045). Yeast cell walls are composed of polysaccharides with an outer layer consisting of strongly glycosylated mannoproteins and an inner layer comprising β-1,6- and β-1,3- glucan chains linked to chitin and mannoproteins [18,20]. Some studies have indicated that the alkaline-insoluble β-1,3-glucan forms a complex structure with many ZEN binding sites, while the alkaline-soluble β-1,6-glucan potentiates this bond [27,28]. These parameters are considered when yeast are used as commercial adsorbent. The wild strains investigated herein display similar adsorption rates to the applied CS, although studies aiming to identify, quantify, and optimise selected microorganisms and their β-glucan concentrations are required for commercial applications. Several studies have focused on gene regulation procedures to optimise yeast strains, leading to a greater number of cell wall binding sites for each specific mycotoxin [22].The adsorption range of mycotoxins was higher at pH 3 compared to pH 6. The literature reports adsorption rates ranging from 20.46 to 54.48%, where yeast cell wall adsorption is favoured under acidic conditions [15,29]. The solubilisation of β-glucans under different pH conditions alters their structure and the conformational energy to create new interactions. pH value variations alter the flexibility of these interactions, leading to higher flexibility under neutral conditions compared to more acidic conditions. The decreasing flexibility noted herein at pH 3.0 increased ZEN adsorption values when compared to neutral and alkaline conditions [28,30]. However, as noted for LL08 adsorption rates at pH 6, interactions are more flexible and effective for AFB1, due to its physicochemical characteristics. This reinforces the need for studies dedicated to cell wall component enhancements to achieve more flexible products capable of adsorbing aflatoxins under different conditions.In the present study, ZEN adsorption rates were higher than AFB1 rates. As the reaction conditions were the same, this difference can be explained based on the physicochemical characteristics of each investigated mycotoxin. The most important physicochemical parameter associated with adsorption is the oil/water partition coefficient (logP), where higher log values result in higher hydrophobicity. The logP values for AFB1 and ZEN are 1.23 and 3.60 respectively [31,32]. As ZEN is highly hydrophobic, it is expected to exhibit greater interaction with strains containing a high number of membrane binding sites, as observed in the present study. Several inorganic-based compounds have been employed to assess aflatoxin adsorptive effects, and the literature indicates that aluminosilicates, bentonites, and zeolites exhibit high in vitro aflatoxin adsorption capabilities [33,34]. Concerning AFB1, adsorption values between 51.94% and 100% for at pH 3 and pH 6 have been reported when comparing the effectiveness of inorganic adsorbents [34-36]. The wild strains investigated herein exhibit equivalent effects to CS, thus considered moderately efficient adsorbents compared to inorganic adsorbents.

Despite the observed effectiveness differences, the use of live microorganisms has been the focus of mycotoxin mitigation in several assessments, due to their probiotic effect and mycotoxin biotransformation associated with microorganism intake [15]. In this regard, recent studies have focused on the optimisation of mycotoxin mitigation in dairy cattle by the association of organic and inorganic adsorbents, to enhance AFB1 removal and concomitantly reduce AFM1 concentrations in milk [37-40].

3.2. Biotransformation Potential

Some studies have highlighted the ability of yeast in metabolising mycotoxins into other compounds. Herein, however, none of the evaluated strains exhibited significant mycotoxin biotransformation potential under the evaluated conditions when compared with the commercial strain.

Higher conversion rates of up to 69-71% have been reported from initial mycotoxin concentrations following longer incubation periods (approximately 48-96 h) [15,41]. The incubation period of the present study, however, was of no more than 24 h, even though the enzymatic synthesis peak of some yeast takes place at longer periods of time, observed by the non-significant relationship between yeast concentrations and pH ($p \geq 0.05$).

Although no significant biotransformations were observed herein, this effect cannot be neglected due to the microorganism's enzymatic complex used for mycotoxin detoxification [21,41].Microorganism mycotoxin biotransformation capabilities are an advantage compared to inorganic adsorbents, as other beneficial effects for target animals are also noted. Furthermore, inorganic compounds can present adverse effects linked to the absorption of micronutrients and other essential compounds [42-44].

The biotransformation assay applied herein may present false-positive results, mainly from chemical hydrolysis due to prolonged MT exposure or the presence of certain components in the reaction medium [45]. Therefore, assays using known derivative standards to correctly evaluate mycotoxin biotransformation by yeast are required [15,46,47].

4. Concluions

The isolation of microorganisms from nature is a promising alternative for the development of new anti-mycotoxin additives with probiotic functions, reinforcing the idea of employing additives of biological origin. In this regard, the
S. cerevisiae strains evaluated herein exhibited excellent adsorptive capacity for AFB1 and ZEN under gastrointestinal conditions. All strains displayed high potential for commercial applications, especially strain LL08, and further studies are required for their optimisation for use in livestock. Although no significant biotransformation was observed, this effect should not be discarded when selecting microorganisms with anti-mycotoxin potential, as many microorganisms can convert mycotoxins into less toxic metabolites in the long term, which is important for the development of new products. Since microorganisms exhibit lower adsorptive capacity compared to inorganic adsorbents, commercial products should be developed by focussing on genomic selection of biotransformation enzymes aiming at mycotoxin detoxification. Eliminating mycotoxins using additives is, however, not enough to ensure feed safety, and mechanisms to control and select quality raw materials should also be applied to improve product quality and animal health.

Author Contributions: Conceptualisation, V.F.M., L.A.P. and F.B.N.K.; Methodology, V.F.M., L.A.P., F.B.N.K., K.M.K., M.A. and L.A.M.K.; Validation, K.M.K., M.A. and L.A.M.K.; Formal analysis, V.F.M. and L.A.M.K.; Investigation, V.F.M., L.A.P., F.B.N.K. and L.A.M.K.; Resources, K.M.K., M.A. and L.A.M.K.; Data curation, V.F.M. and L.A.P.; Writing-original draft preparation, V.F.M., L.A.P. and L.A.M.K.; Writing- review and editing, F.B.N.K., K.M.K., M.A. and L.A.M.K.; Supervision, V.F.M., L.A.P. and L.A.M.K.; Project administration, V.F.M., L.A.P., M.A. and L.A.M.K.; Funding acquisition, L.A.M.K. All authors have read and agreed to the published version of the manuscript.

Funding: This research was funded by FAPERJ-Fundação de Amparo à Pesquisa do Estado do Rio de Janeiro, Process SEI 260003/002531/2021 and SEI 260003/003283/2022.

Institutional Review Board Statement: Not applicable

Informed Consent Statement: Not applicable

Data Availability Statement: Not applicable

Acknowledgements: The authors would like to thank FAPERJ-Fundação de Amparo à Pesquisa do Estado do Rio de Janeiro for scholarship and funding. Thanks are also due to the post-graduation course Higiene Veterinária e Tecnologia de Produtos de Origem Animal of the University Federal Fluminense. the Departamento de Tecnologia de Alimentos at the same university (MTA-UFF), to the Mycology and Mycotoxin Laboratory of the at the Universidade Federal de Minas Gerais (LAMICO-UFMG).

Conflicts of Interest: The authors declare no conflict of interest.

References

1. Peles, F.; Sipos, P.; Kovács, S.; Győri, Z.; Pócsi, I.; Pusztahelyi, T. Biological control and mitigation of aflatoxin contamination in commodities. Toxins **2021**, 1, 104.
2. Yang, C.; Song, G.; Lim, W. Effects of mycotoxin-contaminated feed on farm animals. J. Hazard. Mater. **2020**,

389, 122087.
3. Eskola, M.; Kos, G.; Elliott, C.T.; Hajšlová, J.; Mayar, S.; Krska, R. Worldwide contamination of food-crops with mycotoxins: Validity of the widely cited 'FAO estimate'of 25. Crit. Rev. Food Sci. Nutr. **2020**, 60, 2773- 2789.
4. Bhardwaj, K.; Meneely, J.P.; Haughey, S.A.; Dean, M.; Wall, P.; Zhang, G.; Baker, B.; Elliott, C.T. Risk assessments for the dietary intake aflatoxins in food: A systematic review (2016-2022). Food Control **2023**, 149, 109687.
5. Keller, L.A.M.; Aronovich, M.; Keller, K.M.; Castagna, A.A.; Cavaglieri, L.R.; Rosa, C.A.R. Incidence of Mycotoxins (AFB1 and AFM1) in Feeds and Dairy Farms from Rio de Janeiro State, Brazil. Veterinary. Medicine **2016**, 1, 29-35.
6. Kabak, B. Aflatoxins in foodstuffs: Occurrence and risk assessment in Turkey. J. Food Compost. Anal. **2021**, 96, 103734.
7. Conteçotto, A.C.T.; Pante, G.C.; Castro, J.C.; Souza, A.A.; Lini, R.S.; Romoli, J.C.Z.; Abreu Filho, B.A.; Mikcha, J.M.G.; Mossini, S.A.G.; Machinski Junior, M. Occurrence, exposure evaluation and risk assessment in child population for aflatoxin M1 in dairy products in Brazil. Food Chem. Toxicol. **2021**, 148, 111913.
8. Pitt, J.I. Toxigenic fungi and mycotoxins. Br. Med. J. **2000**, 56, 184-192.
9. Benkerroum, N. Chronic and acute toxicities of aflatoxins: Mechanisms of action. Int. J. Environ. Res. Public Health. **2020**, 17, 423.
10. Wang, Y.; Liu, F.; Zhou, X.; Liu, M.; Zang, H.; Liu, X.; Shan, A.; Feng, X. Alleviation of Oral Exposure to Aflatoxin B1-Induced Renal Dysfunction, Oxidative Stress, and Cell Apoptosis in Mice Kidney by Curcumin. Antioxidants **2022**, 11, 1082.
11. Forsythe, S.J. Microbiology of Food Safety; ArtMed: Porto Alegre, Brazil, 2002; 422p.
12. Rai, A.; Das, M.; Tripathi, A. Occurrence and toxicity of a fusarium mycotoxin, ZENralenone. Crit. Rev. Food Sci. Nutr. **2019**, 60, 2710-2729.
13. Yu, H.; Zhang, J.; Chen, Y.; Zhu, J. Zearalenone and Its Masked Forms in Cereals and Cereal-Derived Products: A Review of the Characteristics, Incidence, and Fate in Food Processing. J. Fungi **2022**, 8, 976.
14. Luo, Y.; Liu, X.; Yuan, L.; Li, J. Complicated interactions between bio-adsorbents and mycotoxins during mycotoxin adsorption: Current research and future prospects. Trends Food Sci. Technol. **2020**, 96, 127-134.
15. Keller, L.A.M.; Abrunhosa, L.; Keller, K.M.; Rosa, C.A.R.; Cavaglieri, L.R.; Venâncio, A. ZENralenone and its derivatives α-ZENralenol and β-ZENralenol decontamination by Saccharomyces cerevisiae strains isolated from bovine forage. Toxins **2015**, 7, 3297-3308.
16. Uyeno, Y.; Shigemori, S.; Shimosato, T. Effect of probiotics/prebiotics on cattle health and productivity.
Microbes Environ. **2015**, 30, 126-132.
17. Cecchini, F.; Morassut, M.; Saiz, J.C.; Moruno, E.G. Anthocyanins enhance yeast's adsorption of Ochratoxin A during the alcoholic fermentation. Eur. Food Res. Technol. **2019**, 245, 309-314.
18. Goncalves, B.L.; Rosim, R.E.; de Oliveira, C.A.F.; Corassin, C.H. The in vitro ability of different Saccharomyces cerevisiae-based products to bind aflatoxin B1. Food Control **2015**, 47, 298-300.
19. Oliveira, A.A.; Keller, K.M.; Deveza, M.V.; Keller, L.A.M.; Dias, E.O.; Martini-

Santos, B.J.; Rosa, C.A.R. Effect of three different anti-mycotoxin additives on broiler chickens exposed to aflatoxin B1. Arch. Med. Vet. **2015**, 47, 175-183.
20. Petruzzi, L.; Corbo, M.R.; Sinigaglia, M.; Bevilacqua, A. Ochratoxin A removal by yeasts after exposure to simulated human gastrointestinal conditions. J. Food Sci. **2016**, 81, 2756-2760.
21. Luo, Y.; Wang, J.G.; Liu, B.; Wang, Z.L.; Yuan, Y.H.; Yue, T.L. Effect of yeast cell morphology, cell wall physical structure and chemical composition on patulin adsorption. PLoS ONE **2015**, 10, e0136045.
22. Luo, Y.; Liu, X.J.; Liu, Y.; Han, Y.Q.; Li, J. Exogenous calcium ions enhance patulin adsorption capability of
Saccharomyces cerevisiae. J. Food Prot. **2019**, 82, 1390-1397.
23. Pfliegler, W.P.; Pusztahelyi, T.; Pócsi, I. Mycotoxins-prevention and decontamination by yeasts. J. Basic Microbiol. **2015**, 55, 805-818.
24. Silva, F.C.; Chalfoun, S.M.; Batista, L.R.; Santos, C.; Lima, N. Polyphasic taxonomy for the identification of
Aspergillus section flavi: A review. Electronic Journal Classroom Focus **2015**, 1, 18-40.
25. Bovo, F.; Corassin, C.H.; Rosim, R.E.; Oliveira, C.A.F. Efficiency of Lactic Acid Bacteria Strains for Decontamination of Aflatoxin M1 in Phosphate Buffered Saline Solution and in Skimmed Milk. Food Bioprocess Technol. **2013**, 6, 2230-2234.
26. Pereyra, C.M.; Gil, S.; Cristofolini, A.; Bonci, M.; Makita, M.; Monge, M.P.; Montenegro, M.A.; Cavaglieri, L.R. The production of yeast cell wall using an agroindustrial waste influences the wall thickness and is implicated on the aflatoxin B1 adsorption process. Food Res. Int. **2018**, 111, 306-313.
27. Aazami, M.H.; Nasri, M.H.F.; Mojtahedi, M.; Mohammadi, S.R. In vitro aflatoxin B1 binding by the cell wall and (1→3)-β-d-glucan of baker's yeast. J. Food Prot. **2018**, 81, 670-676.
28. Hamza, Z.; El-Hashash, M.; Aly, S.; Hathout, A.; Soto, E.; Sabry, B.; Ostroff, G. Preparation and characterisation of yeast cell wall beta-glucan encapsulated humic acid nanoparticles as an enhanced aflatoxin B1 binder. Carbohydr. Polym. **2019**, 203, 185-192.
29. Prapapanpong, J.; Udomkusonsri, P.; Mahavorasirikul, W.; Choochuay, S.; Tansakul, S. In vitro studies on gastrointestinal monogastric and avian models to evaluate the binding efficacy of mycotoxin adsorbents by liquid chromatography-tandem mass spectrometry. J. Adv. Vet. Anim. Res. **2019**, 6, 125-132.
30. Martínez, J.; Hernández-Rodríguez, M.; Méndez-Albores, A.; Téllez-Isaías, G.; Jiménez, E.M.; Nicolás- Vázquez, M.I.; Ruvalcaba, R.M. Computational Studies of Aflatoxin B1 (AFB1): A Review. Toxins **2023**, 15, 153.
31. NCBI. National Centre for Biotechnology Information. PubChem Compound Summary for CID 15558498, Aflatoxin M1. 2022a. Available online: https://pubchem.ncbi.nlm.nih.gov/compound/Aflatoxin-M1 (accessed on 24 March 2023).
32. NCBI. National Centre for Biotechnology Information. PubChem Compound Summary for CID 5281576, ZENralenone. 2022b. Available online: https://pubchem.ncbi.nlm.nih.gov/compound/ZENralenone (accessed on 24 March 2023).
33. Holland, D.M.; Kim, S.W. Mycotoxin Occurrence, Toxicity, and Detoxifying Agents in Pig Production with an Emphasis on Deoxynivalenol. Toxins **2021**, 13, 171.
34. Vila-Donat, P.; Marín, S.; Ramos, A.J. A review of the mycotoxin adsorbing agents,

with an emphasis on their multi-binding capacity, for animal feed decontamination. Food Chem. Toxicol. **2018**, 114, 246-259.
35. Ahn, J.Y.; Kim, J.; Cheong, D.H.; Hong, H.; Jeong, J.Y.; Kim, B.G. An In Vitro Study on the Efficacy of Mycotoxin Sequestering Agents for Aflatoxin B1, Deoxynivalenol, and Zearalenone. Animals **2022**, 12, 333.
36. Yiannikouris, A.; Apajalahti, J.; Kettunen, H.; Ojanperä, S.; Bell, A.N.W.; Keegan, J.D.; Moran, C.A. Efficient Aflatoxin B1 Sequestration by Yeast Cell Wall Extract and Hydrated Sodium Calcium Aluminosilicate Evaluated Using a Multimodal In-Vitro and Ex Vivo Methodology. Toxins **2021**, 13, 24.
37. Jiang, Y.; Ogunade, I.M.; Kim, D.H.; Li, X.; Pech-Cervantes, A.A.; Arriola, K.G.; Oliveira, A.S.; Driver, J.P.; Ferraretto, L.F.; Staples, C.R.; et al. Effect of adding clay with or without a Saccharomyces cerevisiae fermentation product on the health and performance of lactating dairy cows challenged with dietary aflatoxin B1. J. Dairy Sci. **2018**, 101, 3008-3020.
38. Maki, C.R.; Thomas, A.D.; Elmore, S.E.; Romoser, A.A.; Harvey, R.B.; Ramirez, H.A.; Phillips, T.D. Effects of calcium montmorillonite clay and aflatoxin exposure on dry matter intake, milk production, and milk composition. J. Dairy Sci. **2016**, 99, 1039-1046.
39. Rodrigues, R.O.; Rodrigues, R.O.; Ledoux, D.R.; Rottinghaus, G.E.; Borutova, R.; Averkieva, O.; Mcfadden,
T.B. Feed additives containing sequestrant clay minerals and inactivated yeast reduce aflatoxin excretion in milk of dairy cows. J. Dairy Sci. **2019**, 102, 6614-6623.
40. Xiong, J.L.; Wang, Y.M.; Zhou, H.L.; Liu, J.X. Effects of dietary adsorbent on milk aflatoxin M1 content and the health of lactating dairy cows exposed to long-term aflatoxin B1 challenge. J. Dairy Sci. **2018**, 101, 8944- 8953.
41. Li, P.; Su, R.; Yin, R.; Lai, D.; Wang, M.; Liu, Y.; Zhou, L. Detoxification of Mycotoxins through Biotransformation. Toxins **2020**, 12, 121.
42. Barrientos-Velázquez, A.L.; Arteaga, S.; Dixon, J.B.; Deng, Y. The effects of pH, pepsin, cation exchange, and vitamins on aflatoxin adsorption on smectite in simulated gastric fluids. Appl. Clay Sci. **2016**, 120, 17-23.
43. Elliot, C.T.; Connolly, L.; Kolawole, O. Potential adverse effects on animal health and performance caused by the addition of mineral adsorbents to feeds to reduce mycotoxin exposure. Mycotoxin Res. **2019**, 36, 115-126.
44. Wang, G.; Miao, Y.; Sun, Z.; Zheng, S. Simultaneous adsorption of aflatoxin B1 and zearalenone by mono- and di-alkyl cationic surfactants modified montmorillonites. J. Colloid Interface Sci. **2018**, 511, 67-76.
45. Wang, J.; Xie, Y. Review on microbial degradation of ZENralenone and aflatoxins. Grain Oil Sci. Technol. **2020**,3, 117-125.
46. Fruhauf, S.; Novak, B.; Nagl, V.; Hackl, M.; Hartinger, D.; Rainer, V.; Labudová, S.; Adam, G.; Aleschko, M.; Moll, W.D.; et al. Biotransformation of the Mycotoxin ZENralenone to its Metabolites Hydrolyzed ZENralenone (HZEN) and Decarboxylated Hydrolyzed ZENralenone (DHZEN) Diminishes its Estrogenicity In Vitro and In Vivo. Toxins **2019**, 11, 481.
47. Ji, C.; Fan, Y.; Zhao, L. Review on biological degradation of mycotoxins. Anim. Nutr. **2016**, 2, 127-133.

DISCLAIMER/PUBLISHER'S NOTE

SCREENING AND SELECTION OF COMMERCIAL HUMAN PROBIOTIC LACTIC ACID BACTERIA STRAINS FOR BIOTECHNOLOGICAL MILK APPLICATIONS

Submitted to: Microorganisms special issue: Probiotic and Postbiotic Properties of Lactobacillus
Date of submission: 26/09/2023 Date of acceptance:
Date of publication:
The following text complies with the journal's formatting standards as described in the Instructions for Authors section, available at:
https://www.mdpi.com/journal/microrganisms/instructions

Screening and selection of commercial human probiotic lactic acid bacteria strains for biotechnological milk applications

Victor F. Moebus1*, Felipe B.N. Köptcke2, Leonardo A. Pinto3, Tatiana R. Cunha2, Kelly M. Keller4, Eliane T. Mársico5, Luiz A.M. Keller5

[1] PhD student, Veterinary Medicine post-graduation course Veterinary Hygiene and Technology of Products of Animal Origin, Universidade Federal Fluminense, Niterói, Brazil.

[2] Undergraduate student, Pharmacy course, Universidade Federal Fluminense, Niterói, Brazil. felipekoptcke@id.uff.br, tatianareis@id.uff.br

[3] Master's post-graduation course, Plant Biotechnology and Bioprocesses, Federal University of Rio de Janeiro, Rio de Janeiro, RJ. leodeasspinto@gmail.com

[4] Department of Preventive Veterinary Medicine, Veterinary School, University of Coimbra
Federal de Minas Gerais, Belo Horizonte 31270-901, MG, Brazil.
kelly.medvet@gmail.com [5] Associate Professor,
Veterinary Medicine course, Fluminense Federal University,

Niterói, Brazil. etmarsico@id.uff.br, luiz_keller@id.uff.br

* Correspondence: victormoebus@id.uff.br; Tel: +55-21985822408

Abstract: Probiotics can improve host health when ingested in adequate amounts. These microorganisms are highly employed in biotechnology processes, such as fermentation, food technology and detoxification. In this context, the aim of this study was to investigate and characterise the potential of commercial human probiotic strains for biotechnological milk applications. Twenty lactic acid bacteria were investigated concerning kinetic growth curves, gastrointestinal assays, lactose consumption and antimicrobial resistance and bioremoval, as well as aflatoxin M1 (AFM1) bioremoval. Following an initial screening, five Lactobacillus spp. strains were selected as potential biotechnological agents. All achieved the desired concentrations within 24 h and presented antimicrobial resistance, as well as good intestinal adhesion, consisting in viable probiotics with high lactose consumption, thus enabling their application in food technology and organic acid production. Although no AFM1 and antimicrobial biometabolisation was observed, the strains presented AFM1 adsorption rates of up to 65.9%, reducing AFM1 bioavailability in food matrices. The selected strains therefore display the necessary properties for biotechnological milk applications and potential for decontamination, allowing for LAB application in bioremoval technologies, especially in the fermentation processes of low lactose content products.

Keywords: Contaminant bioremoval, AFM1, lactose, antimicrobial.

1. Introduction

Lactic acid bacteria (LAB) are the most predominant probiotic bacteria, comprising the Lactobacillus, Lactococcus, Carnobacterium, Enterococcus, Streptococcus, Pediococcus, Propionibacterium, and Leuconostoc genera, all categorised as "generally recognised as safe" (GRAS) [1,2]. Probiotic bacteria possess the ability to adhere and survive in the gastrointestinal (GI) tract, promoting its stability and protecting this intricate ecosystem, which also facilitates their ability to confer health benefits. The main probiotic bacteria effects comprise the restoration of unbalanced gut microbiota, mitigation or prevention of various gastrointestinal disorders, blood cholesterol reduction, immune system stimulation and the prevention of heart and infectious diseases [3,4].

Lactic acid bacteria share common characteristics, although certain intrinsic species variations, such as genetic and phenotypic variations, are noted. The screening and selection of LAB strains aiming at specific biotechnological applications, such as dairy production, lactic acid production or contaminant removal, are crucial to optimise certain processes and increase final product yields [5,6].

Lactic acid bacteria, as fermentation microorganisms, use carbohydrates, such as lactose in milk, as energy sources and produce a diverse range of bioactive products, including organic acids, bacteriocins, vitamins, exopolysaccharides, gamma-aminobutyric acid, flavouring substances and bioactive peptides. These inherent properties make LAB highly valuable in various dairy product processes, resulting in a wide range of fermented foods with nutraceutical characteristics [7,8,9].

Lactose fermentation by LAB predominantly produces lactic acid. The industrial significance of lactic acid production stems from its wide-ranging applications across various sectors, with food, beverage and pharmaceutical productions and the cosmetics industry as the most noteworthy [10]. The production of compounds obtained through microbial fermentation has emerged as a prominent solution in response to escalating energy demands and environmental challenges, and the advancement of novel techniques employing microorganisms to sustainably generate products through green technology has become imperative [11,12].

The main biotechnological application of LAB in the production of dairy products is in the manufacture of fermented products such as yogurts, fermented milks and kefir, where the use of commercially well-recognized starter cultures are responsible for the characteristic flavour and texture of the products through the incorporation of the metabolites produced [13,14]. The presence of these peptides in a fermented product allows them to be defined as functional foods, capable of providing health benefits beyond their nutritional value, with the selection of strains with desired characteristics being one of the strategies during the development of a dairy derivative [15].

In the last few decades, LAB-based biological approaches, mainly biotransformation and adsorption processes, have become prominent as promising milk decontamination methods. Lactic acid bacteria, therefore, offer potential solutions for ensuring the safety and quality of dairy products affected by different contaminants [16,17]. Several aflatoxin M1 (AFM1) LAB-removal assessments are available, with positive results indicating reduced mycotoxin bioavailability. Antimicrobial milk residue removal employing LAB has, however, not been widely evaluated due to antimicrobial decreases during heat treatments. The risks associated with the presence of these contaminants, however, indicates the need for studies focused on the development of methodologies capable of reducing antimicrobial bioavailability, ensuring the safety and quality of milk and dairy products [18,19].

Despite the use of starter cultures for the production of dairy products is well described, it is necessary to discover and select new LAB, such as those used as human probiotics, capable of combining the potential for removing and mitigating contaminants with the production of bioactive compounds. capable of promoting the improvement of the final quality of the product while maintaining beneficial effects for the consumer [14].

In this context, the aim of this study was to screen and select LAB strains used as commercial human probiotics for biotechnological milk applications concerning potential contaminant bioremoval.

2. Material and Methods

*2.1*Bacterial strains and growth conditions

Twenty freeze-dried LAB strains from commercial human probiotics (Table 1) were investigated concerning potential biotechnological milk applications at the Federal Fluminense University Laboratory of Microbiological Control of Products of Animal Origin (LCMPOA), Rio de Janeiro, Brazil. The strains were reactivated prior to the experiments using Man Rugosa Sharpe (MRS) broth (KASVI®, São José dos Pinhais, BR), for 24 h at 37 °C [20].

Table 1 Species, strain identity and commercial source.

Species	Strain identity	Commercial Source
Lactobacillus acidophilus	Acidophilus LYO®	Fermentech, Tatuapé, Brazil
Lactobacillus acidophilus	LA-5®	Chr. Hansen, Hørsholm, Denmark
Lactobacillus acidophilus	ACID GB®	Gabbia Biotechnology and Development, Barra Velha, Brazil
Lactobacillus rhamnosus	GR-1®	Chr. Hansen, Hørsholm, Denmark
Lactobacillus rhamnosus	LGG®	Chr. Hansen, Hørsholm, Denmark
Lactobacillus rhamnosus	Rhamnosus LYO®	Fermentech, Tatuapé, Brazil
Lactobacillus rhamnosus	CRL 1505®	Centro Sperimentale Del Latte, Zelo Buon Persico, Italy
Lactobacillus bulgaricus	LB 340 LYO®	Fermentech, Tatuapé, Brazil
Lactobacillus bulgaricus	LB-G40®	BioGrowing Co., Shanghai, China
Lactobacillus bulgaricus	BULG GB®	Gabbia Biotechnology and Development, Barra Velha, Brazil
Streptococcus thermophilus	TH-4®	Chr. Hansen, Hørsholm, Denmark
Streptococcus thermophilus	TA 40 LYO®	Chr. Hansen, Hørsholm, Denmark
Bifdobacterium lactis	Bifido LYO®	Fermentech, Tatuapé, Brazil
Bifdobacterium lactis	LACT GB®	Gabbia Biotechnology and Development, Barra Velha, Brazil
Lactobacillus casei	L. CASEI 431®	Chr. Hansen, Hørsholm, Denmark
Lactobacillus paracasei	NTU 101®	Centro Sperimentale Del Latte, Zelo Buon Persico, Italy
Lactobacillus casei	CAS GB®	Gabbia Biotechnology and Development, Barra Velha, Brazil
Lactobacillus reuteri	RC-14®	Chr. Hansen, Hørsholm, Denmark
Lactobacillus gasseri	LG08®	Wecare Biotechnology Co, Jiangsu,China
Lactobacillus gasseri	GASS GB®	Gabbia Biotechnology and Development, Barra Velha, Brazil

2.2 Screening methods and selection parameters

3.1.1 Kinetic growth curve construction

The kinetic growth curves for each strain were constructed according to Malik et al. [21]. Briefly, the reconstituted cells were inoculated in 300 mL of MRS to obtain an initial concentration of 10^{6} CFU mL-1 and incubated at 37 °C for 48 hours under anaerobic conditions. Viable cell counts and optical density at 600 nm (OD600) were verified at 0, 2, 4, 8, 12, 18, 24, 36 and 48 hours according to the Compendium of Methods for the Microbiological Examination of Foods [22]. The OD600 values were verified using a Shimazdu UV-PROBE 1800 UV/VIS spectrophotometer (SHIMADZU ®, Kyoto, Japan) at 600 nm. The samples were diluted to conform to Lambert-Beer's Law [23] and all experiments were performed in duplicate and repeated three times.

2.2.2 Bacterial tolerance under gastrointestinal conditions

Bacterial tolerance to different environments was assessed in an in vitro model simulating different acidic conditions. Bacterial growth was evaluated using phosphate and citrate buffers at pH 2 and pH 3 for acidic conditions and pH 7 and pH 6 for neutral conditions, respectively.

All solutions were prepared fresh and sterilised on the same day as the experiments. For the tolerance assay, 1 mL of a 10^{8} CFU mL^{-1} bacterial suspension inoculum was added to 900 μL of each buffer and incubated for 24 hours at 37 °C under anaerobic conditions. The number of viable cells was then determined through serial dilutions plated on MRS agar and incubated anaerobically at 37 °C for 96 hours. All assays were performed in duplicate and repeated three times [22].

2.2.3 Bacterial adhesion assay

Bacterial adhesion to human colon cells was assessed in an in vitro model [24] employing Caco-2, HT-29 and LS 174T human colon adenocarcinoma cell lines. The bacterial inoculums were added at a final concentration of 10^{8} CFU in 1 mL of the cell media per well. Following 1 hour of incubation, the cell layers were washed three times with PBS to remove non-adherent bacteria and lysed with 0.1% Triton-X100 in PBS (Sigma-Aldrich, St. Louis, USA). The viability of the adhered cells was confirmed by serial dilutions, plating on MRS agar and counting following incubation under anaerobic condition for 96 hours at 37 °C. All assays were performed in duplicate and repeated three times [22].

2.2.4 Antimicrobial resistance assays

The macrodilution method, with modifications, was used for the antimicrobial (AM) resistance assays [25]. The tested antimicrobials and their concentrations were chosen according to Brazilian regulations [26], as follows: streptomycin (3200, 1600, 800, 400, 200, 100, 50, 25, 12.5 and 6.25 μ L^{-1}), penicillin G (64, 32, 16, 8, 4, 2, 1, 0.5, 0.25 and 0.125 μ L^{-1});tetracycline (1600, 800, 400, 200, 100, 50, 25, 12.5, 6.25 and 3.125 μ L^{-1}) and azithromycin (640, 320, 160, 80, 40, 20, 10, 5, 2.5 and 1.25 μ L^{-1}). All solutions were prepared in a pH 7.2 phosphate buffer solution (PBS).For the resistance tests, tubes containing

2 mL of Mueller-Hinton broth (KASVI®, São José dos Pinhais, Brazil), 1 mL of the antimicrobial solutions at different concentrations to reach the desired concentrations and 1 mL of each bacterial inoculum to obtain 10^8 CFU mL^{-1} were incubated at 37 °C for 48 hours under anaerobic conditions. A positive control was prepared by adding 1 mL of peptone saline solution 1% instead of the antimicrobial solution and a negative control was prepared by adding 1 mL of peptone saline solution 1% in place of the bacterial inoculum, maintaining a final volume of 4 mL in each tube.Viable cell counts after incubation were conducted through plate counts using MRS agar (KASVI®, São José dos Pinhais, Br) following incubation at 37 ° C for 96 hours under anaerobic conditions.

2.3 Biotechnological characterisation

2.3.1 Lactose consumption

Lactose consumption was evaluated in vitro by adding each bacterial inoculum at 10^8 CFU mL^{-1} to 10 mL of lactose broth (5 g L^{-1}) at an initial pH value of 7.0 (Difco®, Becton Dickinson, New Jersey, USA). The samples were then incubated at 37°C for 24 hours under anaerobic conditions . Lactose quantification was performed as described by Ni et al. [27] for reducing sugars. The final pH values following the incubation were determined using a pHmeter ion phs-3e apparatus (Satra®, Kettering, United Kingdom).

2.3.2 Aflatoxin M1 quantification

An HPLC system comprising a Shimadzu LC-20AD pump, a Shimadzu SIL-10AF autosampler, Shimadzu CTO-20A furnace and a Shimadzu RF-20A fluorescence detector (all from Shimadzu®, Kyoto, Japan), was employed for the AFM1 determinations. Samples were separated using a Spherisorb column (150mm × 4.6mm,5 μm) (Waters®, Massachusetts, USA) and AFM1 quantifications were performed according to Shundo et al. [28]. The following chromatographic conditions were employed: an aqueous 2% acetic acid:acetonitrile:methanol solution (40:35:25, v/v) as the mobile phase at 1.0 mL min^{-1}, injection volume of 50 μL and oven temperature of 40°C. Fluorescence detection was performed at an excitation wavelength of 365 nm and an emission wavelength of 445 nm. The retention time for AFM1 was 2 minutes under these conditions. The calibration curve comprised 50 μL injections corresponding to 0.01; 0.05; 0.5; 1.0; 2.0; 4.0 ng of the stock solution. The determined Limit of Detection (LOD) was 0.01 ng mL^{-1} and the Limit of Quantification (LOQ), 0.05 ng mL^{-1}.

2.3.3 Aflatoxin M1 bioremoval

2.3.3.1 Aflatoxin M1 adsorption

The in vitro evaluation of the adsorption potential of the investigated probiotic strains was performed as described by Keller et al. [29] and Moebus et al. [30]. Assays were conducted with both viable and inactivated cells. The strains were first cultivated in 250 mL MRS at 37 °C under anaerobic conditions to obtain 10^8 CFU mL^{-1}. The culture broths were then centrifuged at 1,500 rpm for 15 min, discarding the supernatants, and the pellets washed twice with PBS to minimise interferences. For inactivated cells, the strains were heated at 100 °C for one hour after an incubation period, and precipitation was performed as described for viable cells. The bacterial pellets were resuspended in PBS and 1.5 mL were transferred to centrifuge microtubes, supplemented with 2 μL^{-1} of AFM1 and incubated at 37 °C for 24 hours. Following the different incubation periods (8, 12, 18, and 24 hours), the solutions were centrifuged at 1,500 rpm for 15 min and the supernatants transferred to glass vials for the HPLC-FL detections.

The AFM1 adsorption percentages were calculated using Equation 1:

AFM1 concentration in *the* supernatant AFM1 concentration in *the* positive control

2.3.3.2 Aflatoxin M1 biotransformation

For the desorption assay, 1.5 mL PBS were added to each microtube and the samples resuspended to release AFM1 molecules bound to cell structures for desorption. Following this step, the tubes were again centrifuged and the supernatants collected into new vials for HPLC quantification. Biotransformed AFM1 concentrations were determined by the difference between the adsorption and desorption assay values, according to Equation 2:

AFM1b = AFM1i - AFM1ads - AFM1des(2)

Where AFM1b is the biotransformed mycotoxin concentration following the incubation period, AFM1i is the initial mycotoxin percentage present in the assay AFM1ads, is the adsorbed mycotoxin concentration and AFM1des is the desorbed free mycotoxin concentration [29,30].

2.3.4 Antimicrobial bioremoval

2.3.4.1 Antimicrobial adsorption

The in vitro antimicrobial adsorption potential evaluations of the probiotic strains were performed as described by Keller et al. [29], Moebus et al. [30] and Massoud et al [31]. The assays were conducted with both viable and inactivated cells.

The bacterial pellets were resuspended in PBS and 1.5 mL aliquots were transferred to centrifuge microtubes, supplemented with streptomycin 400, 200, 100 $\mu\ L^{-1}$; penicillin G 8, 4, 2 $\mu\ L^{-1}$; tetracycline 200, 100, 50 $\mu\ L^{-1}$ and azithromycin 80, 40, 20 $\mu\ L^{-1}$ and incubated at 37 °C for 24 hours. The samples were centrifuged following the incubation period and the supernatants purified employing SPE RIDA® C18 columns (r-Biopharm®, Marshall, USA). Antimicrobial quantifications were performed according to the Brazilian Pharmacopeia for the referred antimicrobials [32].Antimicrobial adsorption percentages were calculated using Equation3:

AM concentration in *the* supernatant AM concentration in *the* positive control

2.3.6 Antimicrobial (AM) biotransformation

For the desorption assay, 1.5 mL of PBS were added to each microtube and the samples resuspended to release antimicrobials bound to cell structures by desorption. Subsequently, the tubes were centrifuged again and the supernatants collected for quantification. Biotransformed AM concentrations were determined by the difference between adsorption and desorption assay values, according to Equation 4:

AMb = AMi - AMads – Amdes (4)

Where AMb is the biotransformed antimicrobial concentration following the incubation period, AMi is the initial mycotoxin percentage present in the assay AMads, is the adsorbed antimicrobial concentration and AMdes is the desorbed free antimicrobial concentration [29,30].

2.4 Statistical analyses

Results are expressed as the means and standard error means (SEM) of three experiments with triplicate determinations. All statistical analyses were performed using the Infostat version 2020 software (Infostat Software, Córdoba, Argentina). In vitro transit tolerance, kinetic growth and AFM1 and antimicrobial bioremoval were analysed using Spearman's correlation analysis. The antimicrobial resistance assay results were analysed using Pearson's correlation test and lactose consumption was analysed by a one-way analysis of variance (ANOVA) applying Tukey's post-hoc test. Differences were considered statistically significant at $p \geq 0.05$.

3. Results and discussion

Three crucial parameters were defined to establish screening protocols, namely the capability to attain a 108 CFU mL-1 concentration within 24 hours, assessed through the growth kinetics, growth capacity under various incubation systems, mainly cellular concentration at specified time intervals and resistance against antimicrobial agents. Following growth kinetics assessments under different incubation regimes, only five strains were found to reach 108 CFU mL-1 in 24 hours, namely L. acidophilus ACID GB®, L. bulgaricus LB-G40®, L. casei NTU 101®, L. gasseri LG08®,L. rhamnosus CRL 1505®. The findings of the parameter selection assays and Potential biotechnological application results are described below.

3.1 Screening protocols and parameter selection

3.1.1 Kinetic growth curves

The kinetic growth curves of the five selected Lactobacillus spp. strains are depicted in Figures 1, 2, 3, 4, and 5, and the kinetic parameters and correlation coefficients are presented in Table 2.

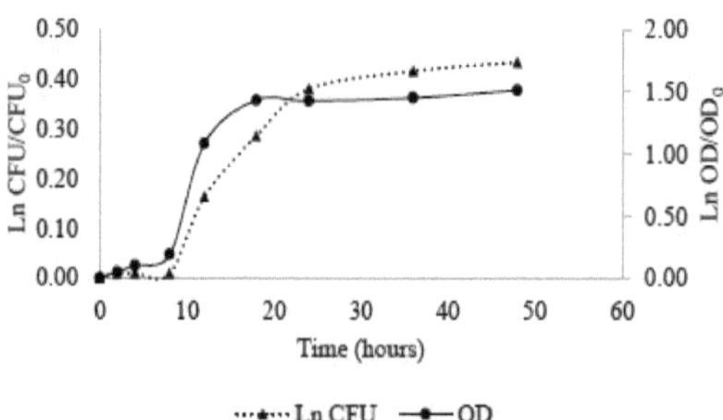

Figure 1: Kinetic growth curve for L. acidophilus ACID GB®

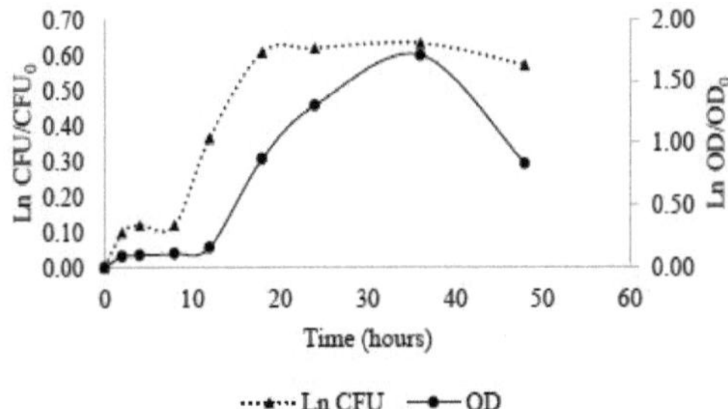

Figure 2: Kinetic growth curve for L. bulgaricus LB-G40®

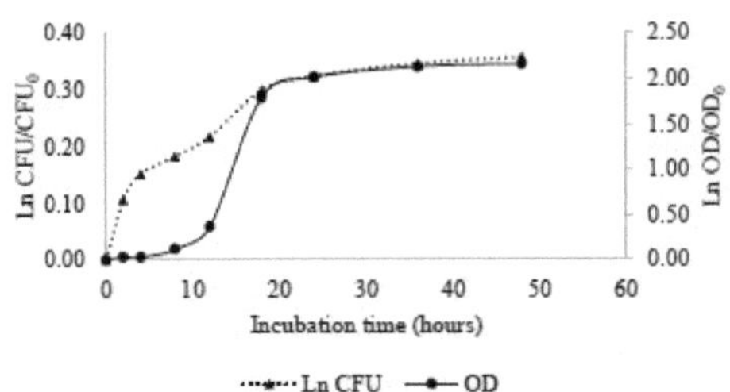

Figure 3: Kinetic growth curve for L. casei NTU 101

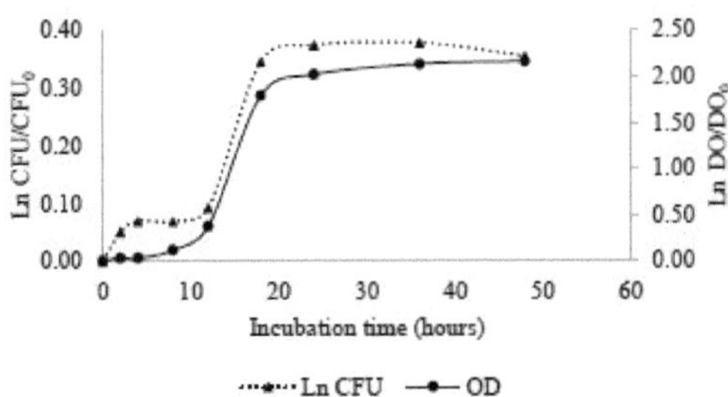

Figure 4: Kinetic growth curve for L. gasseri LG08®

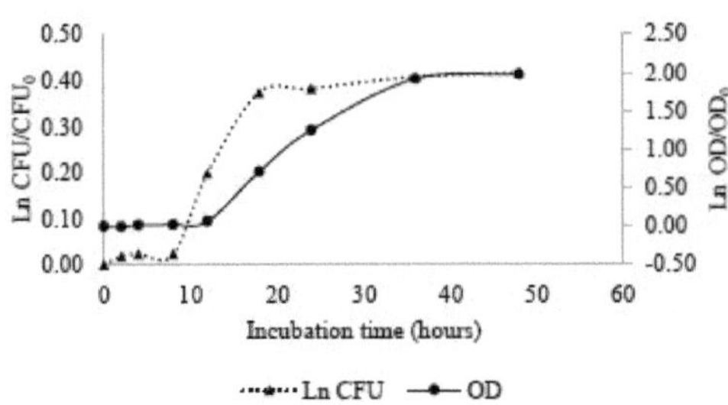

Figure 5: Kinetic growth curve for L. rhamnosus CRL 1505®

Table 2: Kinetic parameters and correlation coefficient of selected lactic acid bacteria

Lactobacillus strain	µ max1 (h-1)	gt2 (h)	r2 *	P value
L. acidophilus ACID GB®	0.0206	33.65	0.93	<0.05
L. bulgaricus LB-G40®	0.0405	17.11	0.99	<0.05
L. casei NTU 101	0.0137	50.59	0.97	<0.05
L. gasseri LG08®	0.0423	16.39	0,90	<0.05
L. rhamnosus CRL 1505®	0.0291	23.82	0.92	<0.05

[1] Maximum growth speed, [2] Generation time, * Spearman's correlation coefficient.

Five of the twenty LAB strains exhibited viable cell counts of 10^8 CFU mL^{-1} after 24 hours of incubation. High correlation coefficients (r^2) indicate strong associations between viable cell counts and OD600 measure. These findings highlight the potential application of these strains as probiotics and in diverse biotechnological processes, allowing for the quick preparation of standard inoculums in short periods of time.

The generation time (gt) of each strain also provides important information on microbial growth.. L. gasseri presented the shortest generation time among the evaluated strains, requiring only 16.39 hours to double the number of viable cells, while L. casei required 50.59 hours.

Kinect growth rates play an important role in the selection of microorganisms as probiotics. As per FAO and Brazilian laws, concentrations ranging from 8.0 to 9.0 log10CFU mL^{-1} are required for probiotics to exert their beneficial effects following ingestion [33,34]. Probiotics with high growth rates and lower generation times have had a notable impact on the probiotic industry, enabling accelerated production due to shorter multiplication times, complemented by effective drying systems [35].

With regard to biotechnological applications, growth rates are widely regarded as the most crucial microorganism selection factor [23,36]. The literature, for example, consistently highlights the use of concentrations of over 7.0 log10CFU mL^{-1} for various biotechnological applications, including dairy product production and the removal of several contaminants, including AFM1, antimicrobials, and metals, from various matrices [26,37,38].

As seen in Figures 1 to 5 above, the microorganisms showed an excellent incubation time of between 25-35 hours to reach the concentration required for various biotechnological applications. Being excellent candidates for future studies on their application for manufacturing fermented dairy products.

The production of fermented dairy products is directly related to the generation time and maximum growth speed of the microorganisms used in bioprocesses. The growth parameters of LAB, together with the production of metabolites of interest, have been the focus of research over the years to obtain starter cultures that meet production demands. LAB strains commonly used as starter cultures show optimal growth in milk, although they may have limited biosynthetic abilities compared to wild strains. Strains used in bioprocesses for the production of dairy products have been selected over the years, obtaining microorganisms with metabolic simplification due to the loss of ancestral genes [13,14].

The current viable cell count method applied to LAB, however, displays certain limitations that can affect its practical applications. For example, the standard counting method for anaerobically viable cells requires an incubation time of 96 hours [22], which can result in cell concentration disparities between the end of the propagation time and confirmative cell counts. Alternative methodologies, such as the determination of optical density through spectrophotometric assessments (OD600), are, therefore, required to quickly assess cell concentrations and verify inoculum concentrations at the end of the incubation period [23].

Highly significant correlations between OD600 and viable cell count of the selected Lactobacillus strains were observed ($P<0.05$). This provides quick and accurate cell concentration assessments during all incubation stages, allowing for the production of standardised inoculums suitable for varied applications while minimising the loss of viable cells, required for confirmation [26,35,37,38].The observed correlations hold notable implications for practical applications, enabling the efficient and accurate execution of the proposed screening assays, also presenting opportunities for novel applications beyond initial intentions. Trials simultaneously evaluating viable cell counts and OD600 are thus paramount for microorganism selection in diverse technological applications, since knowledge about bacterial concentration influences several stages of bioprocesses, potentially altering production time and sensory characteristics of the dairy product.

3.1.2 Bacterial tolerance under gastrointestinal conditions and bacterial adhesion assays

All selected LAB strains were tolerant to gastrointestinal conditions, displaying high adhesion rates and, thus, establishing them as excellent probiotics for human use. According to FAO guidelines, probiotic microorganisms must display gastrointestinal tolerance to withstand gastric juices and bile salts, as well as the capacity to adhere to the intestinal mucosa in order to exert their beneficial probiotic effects [33].The presence of these probiotics in the intestinal environment modulates the immune system and enhances consumer health. They also prevent various infections and act against multi-resistant strains through competitive inhibition and the production of antimicrobial peptides [2,39,40].

3.1.3 Antimicrobial resistance assays

Antimicrobial agents have been used for decades in livestock production for the prevention, treatment, and prophylaxis of common pathologies. When the recommended withdrawal period is not adhered to, however, antimicrobial residues may be passed on to milk. This may lead to negative effects concerning dairy products, leading to failures in the production of fermented items [41,42]. These issues occur due to partial or complete LAB multiplication inhibition, resulting in delayed acid production by these bacteria, which may, in turn, lead to premature fermentation by pathogenic microorganisms and negatively affect the sensory characteristics of the final products [42,43]. Antimicrobial residues in food are also associated with various toxic outcomes in humans, including allergies, immunopathological effects, carcinogenicity, mutagenicity, nephropathy, hepatotoxicity, reproductive disorders, bone marrow toxicity, and anaphylactic shock [44]. In this context, the selection of LAB strains resistant to antimicrobials commonly used in livestock production is required for biotechnological applications to contaminated milk. Tables 3, 3, 5 and 6 indicate the antimicrobial resistance of selected Lactobacillus probiotic strains investigated herein.

Table 3: Streptomycin Lactobacillus strain resistance.

Streptomycin concentration (μ L-1)

	6.25 12.5 25 50	200*	400 8		1600	3200	C+	C-
L. acidophilus ACID GB®	+ + + +	+ +	+	+	+	+	8.74 ± 0.31	8.68 ± 0.30
L. bulgaricus LB-G40®	+ + + +	+ +	+	+	-	-	8.27 ± 0.28	8.61 ± 0.29
L. casei NTU 101	+ + + +	+ +	+	+	+	+	8.23 ± 0.30	8.64 ± 0.34
L. gasseri LG08®	+ + + +	+ +	+	+	+	+	7.72 ± 0.33	8.65 ± 0.31
L. rhamnosus CRL 1505®	+ + + +	+ +	+ +		+	+	10.02 ± 0.34	8.66 ± 0.27

+: Growth in reaction medium; -: No growth in the reaction medium. *Maximum permissible limit established by the Brazilian legislation [26]; C+: Viable cell counts at higher concentrations (log10CFU mL-1); C-: Negative control (log10CFU mL-1); P value according to Pearson's correlation <0.05.

Table 4: Penicillin G Lactobacillus strain resistance.

Penicillin G concentration (μ L-1)

	0.125	0.25	0.50	1	2	4	8	1	32	64	C+	C-
L. acidophilus ACID GB®	+	+	+	+	+	+	+	+	+	+	8.88± 0.32	8.68 ± 0.30
L. bulgaricus LB-G40®	+	+	+	+	+	+	+	+	+	+	7.78± 0.34	8.61 ± 0.29
L. casei NTU 101	+	+	+	+	+	+	+	+	+	+	8.43± 0.29	8.64 ± 0.34
L. gasseri LG08®	+	+	+	+	+	+	+	+	+	+	7.80± 0.30	8.65 ± 0.31
L. rhamnosus CRL 1505®	+	+	+	+	+	+	+	+	+	+	8.28± 0.31	8.66 ± 0.27

+: Growth in reaction medium; -: No growth in the reaction medium. * Maximum permissible limit established by the Brazilian legislation [26]; C+: Viable cell counts at higher concentrations (CFU mL-1); C-: Negative control (CFU mL-1); P value according to Pearson's correlation < 0.05.

Table 5: Tetracycline G Lactobacillus strain resistance.

Tetracycline concentration (μ L-1)

	3.125	25	12.5	25	50	100*	200		800	1600	C+	C-
L. acidophilus ACID GB®	+	+	+	+	+	+	+	+	+	+	8.29 ± 0.33	8.68 ± 0.30
L. bulgaricus LB-G40®	+	+	+	+	+	+	+	+	+	+	7.68 ± 0.27	8.61 ± 0.29
L. casei NTU 101	+	+	+	+	+	+	+	+	+	+	8.08 ± 0.29	8.64 ± 0.34
L. gasseri LG08®	+	+	+	+	+	+	+	+	+	+	7.12 ± 0.32	8.65 ± 0.31
L. rhamnosus CRL 1505®	+	+	+	+	+	+	+	+	+	+	8.68 ± 0.30	8.66 ± 0.27

+: Growth in reaction medium; -: No growth in the reaction medium. * Maximum permissible limit established by the Brazilian legislation [26]; C+: Viable cell counts at higher concentrations (log10CFU mL-1); C-: Negative control (log10CFU mL-1); P value according to Pearson's correlation <0.05.

Table 6: Azithromycin G Lactobacillus strain resistance

Azithromycin concentration (μ L-1)

	1.25	2.5	5	10	20	40*	80	160	320	640	C+	C-
L. acidophilus ACID GB®	+	+	+	+	+	+	+	+	+	+	8.46± 0.32	8.68 ± 0.30
L. bulgaricus LB-G40®	+	+	+	+	+	+	+	+	+	+	7.80± 0.35	8.61 ± 0.29
L. casei NTU 101	+	+	+	+	+	+	+	+	+	+	8.59± 0.28	8.64 ± 0.34
L. gasseri LG08®	+	+	+	+	+	+	+	+	+	+	7.92± 0.29	8.65 ± 0.31
L. rhamnosus CRL 1505®	+	+	+	+	+	+	+	+	+	+	9.29± 0.31	8.66 ± 0.27

+: Growth in reaction medium; -: No growth in the reaction medium. * Maximum permissible limit established by the Brazilian legislation [26]; C+: Viable cell counts at higher concentrations (log10CFU mL-1); C-: Negative control (log10CFU mL-1); P value according to Pearson's correlation <0.05.

According to CLSI guidelines, Lactobacillus spp. display an intrinsic resistance to vancomycin and other DNA and cell wall synthesis inhibitors, such as ampicillin and cephalosporins [25,45,46], corroborating the results observed herein. Based on the maximum limits established by Brazilian legislation for antimicrobial residues in milk, of 200 μ L-1 for streptomycin, 4 μ L-1 for penicillin G, 100 μ L-1 for tetracycline, and 40 μ L-1for azithromycin [26], the selectedLactobacillus strains were considered asindeed suitable for use in milk biotechnological processes, even in the presence of a wide range of antimicrobials. The selected LAB strains presented resistance to all evaluated antimicrobials. L. bulgaricus is the only exception, being susceptible to higher streptomycin concentrations, enabling its application in matrices containing high concentrations of different antimicrobial agents. The observed viable cell counts in C+ demonstrate the ability of the selected LAB to maintain the necessary concentrations for their application as probiotics in the presence of antimicrobials agents. When tested against streptomycin and azithromycin, some strains, such as L. rhamnosus and L. gasseri respectively, exhibited greater growth compared to the negative control. The decreased viable cell counts observed for

L. gasseri and L. bulgaricus in certain antimicrobial solutions reinforces the need to apply higher inoculum concentrations to ensure the minimum cell concentrations required for biotechnological applications [26,37,38].

The antimicrobial resistance exhibited by the investigated strains offers the potential for conducting tests focused on the bioremoval of antimicrobial agents in diverse matrices, particularly in milk, Although the presence of drug residues in milk is not recommended, studies aimed at mitigating and reducing the bioavailability of contaminants by fermenting microorganisms is necessary to ensure consumer safety and guarantee the application in dairy product industry biotechnological processes [42].

3.2 Biotechnological characterisation

3.2.1 Lactose consumption

Lactose, as the primary sugar in milk, serves as a substrate for LAB growth through fermentation. The fermentation process is responsible for the production of diverse fermented dairy products worldwide, including yoghurt, cheese, fermented milk, acidophilus milk, and kefir [10,47,48,49,50]. During LAB fermentation, the primary metabolism utilises glucose derived from lactose enzymatic hydrolysis, producing multiple byproducts. Lactic acid is the main product formed during fermentation, followed by other organic acids, bacteriocins, and bioactive peptides, all of which contribute to the sensory characteristics, preservation, and safety of fermented dairy products [10,51]. Considering the wide range of biotechnological applications currently available, lactose consumption assessments by LAB are essential for microorganism selection aiming at fermentation processes. The lactose consumption rates and final pH values of the selected Lactobacillus spp. strains investigated herein are presented in Table 7.

Table 7: Lactose consumption rates of Lactobacillus strains and final pH values after incubation.

Lactobacillus strainLactose consumption (%)pH

99.78 ± 0.24A3.74

L. acidophilus ACID GB®

L. bulgaricus LB-G40®

L. casei NTU 101

L. gasseri LG08®

99.70 ± 0.03 A 3.54

99.70 ± 0.06 A 4.63

99.79 ± 0.02 A 4.77

L. rhamnosus CRL 1505® 99.84 ± 0.07[a] 3.60

A: Means followed by the same letter are not significantly different (p >0.05).

The five selected Lactobacillus spp. strains presented excellent lactose consumption rates of over 99% in the applied culture medium. While none of the strains were specifically designed for industrial and dairy product applications, their remarkable ability to consume lactose makes them highly promising for many bioprocesses. The pH assessment indicated the production of various acidic compounds during the fermentation process, evidenced by a decrease in the initial pH broth value (7.0). The final pH values, however, differed among the investigated strains, despite similar lactose consumption levels, suggesting variations in the types and amounts of acidic compounds generated by each strain. Such metabolic differences may lead to implications concerning sensory properties, overall fermented product characteristics, and byproduct yields [10,47,48,49,50]. The dairy industry has actively sought different uses for fermentation process byproducts. Extensive research has been conducted on developing nutraceutical products that incorporate bioactive peptides, as these demonstrate significant consumer health effects, particularly on the cardiovascular and immune systems, displaying anti-inflammatory and antioxidant activities and the ability to modulate insulin secretion [8,9]. Lactose, as a dairy product component, can induce clinical conditions in susceptible individuals, such as allergic diseases and lactose intolerance. The incorporation of LAB, coupled with lactase enzymes in dairy products, thus comprises a valuable strategy in the production of lactose-free products targeted at these consumers. Moreover, probiotic consumption can also assist in managing these clinical conditions by fermenting ingested lactose and alleviating lactose intolerance symptoms [50].Lactic acid can be applied in various industries in both pure and racemic forms. The isolation of lactic acid from other compounds, however, requires several separation and purification steps [10,51].The production of lactic acid by BAL is also important in food preservation, either by inhibiting the growth of pathogenic microorganisms or by preventing the formation of compounds that are harmful to health. Some microorganisms have the ability to produce biogenic amines through the presence of specific enzymes, such as amino acid decarboxylase, biogenic amine oxidase, and biogenic amine dehydrogenase [52,53]. The reduction in pH caused by the production of lactic acid has the ability to inhibit these enzymes, preventing the formation of toxic metabolites. This reduction is still associated with the control o f acrylamide production in foods by reducing the concentration of precursors, such as asparagine and reducing sugar as demonstrated by Bartkiene et al. [54,55].The Lactobacillus spp. strains investigated herein demonstrate potential as lactic acid producers, although further studies to identify and quantify the byproducts produced by each strain are required, in order to obtain higher industrial yields. The observed lactose consumption and acid production by the selected strains, therefore, not only offers potential for industrial applications in dairy product production but also provide an opportunity to obtain lactic acid sustainably. This opens up possibilities for their use in processes aimed specifically at lactic acid production, broadening their industrial applicability beyond the dairy sector.

3.4.2 Aflatoxin M1 bioremoval

The global occurrence of AFM1 in milk and dairy products has emerged as a critical issue, due to the high consumption of these foodstuffs worldwide, as well as the potential human health risks of AFM1, especially in immunocompromised individuals. This compound exhibits carcinogenic, cytotoxic, teratogenic, and genotoxic properties, and prolonged exposure to AFM1 leads to chronic conditions such as immunosuppression, hepatocarcinoma, and stunted growth in children [38,56,57].

Due to its thermoresistant properties, AFM1 can withstand typical dairy product heat treatments. This has led to the development of various AFM1 milk removal methodologies. Most employ microorganisms with the ability to reduce AFM1 bioavailability and biotransformation and adsorption mechanisms, aiming at mitigating the potential health risks associated with AFM1 contamination in dairy products [38]. The adsorption capacity assay of selected LAB strains after 24 hours of incubation are presented in Figure 6.

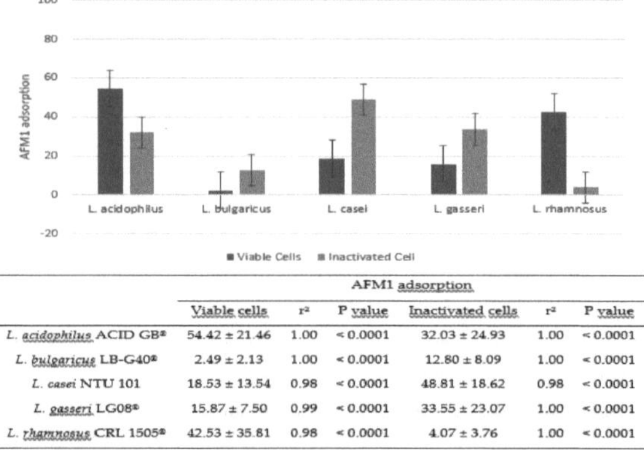

	AFM1 adsorption					
	Viable cells	r^2	P value	Inactivated cells	r^2	P value
L. acidophilus ACID GB®	54.42 ± 21.46	1.00	< 0.0001	32.03 ± 24.93	1.00	< 0.0001
L. bulgaricus LB-G40®	2.49 ± 2.13	1.00	< 0.0001	12.80 ± 8.09	1.00	< 0.0001
L. casei NTU 101	18.53 ± 13.54	0.98	< 0.0001	48.81 ± 18.62	0.98	< 0.0001
L. gasseri LG08®	15.87 ± 7.50	0.99	< 0.0001	33.55 ± 23.07	1.00	< 0.0001
L. rhamnosus CRL 1505®	42.53 ± 35.81	0.98	< 0.0001	4.07 ± 3.76	1.00	< 0.0001

Figure 6. AFM1 adsorption of viable and inactivated cells and statistical comparisons to the total AFM1 bioremoved in the assay. r2: Spearman correlation coefficient.

The adsorptive capacity of viable cells ranged from 2.13% to 54.42% (0.04 to 1.09 μ L^{-1}), while the adsorption of inactive cells ranged from 8.09% to 48.81% (0.16 to 0.98 μ L^{-1}). Strong Spearman correlations (r^2) were noted between adsorbed AFM1 and the total concentration of bioremoved AFM1 in both cell conditions. This indicates that the

The main AFM1 removal mechanism is adsorption. The obtained adsorption ranges are consistent with literature reports between 8% and 100%, corroborating the observed adsorption patterns [17,58,51,59,60]. Similar to yeast cell wall adsorption, LAB are also able to adsorb mycotoxins through hydrophobic and electrostatic interactions. These interactions take place between mycotoxin molecules and the binding sites created by LAB cell constituents, including weak hydrogen bonds and van der Waals bonds [61]. Lactic acid bacteria cell walls are composed primarily of peptidoglycans, which are adorned with teichoic and lipoteichoic acids, a proteinaceous S layer, and polysaccharides. Several studies have indicated that both peptidoglycan and polysaccharides play a role in mycotoxin binding processes [62,63]. A significant difference was noted here concerning adsorption rates for inactivated cells compared to viable cells. Thermal inactivation processes disrupt cell wall structures, exposing cell wall components and enabling enhanced AFM1 interactions. The enhanced adsorption noted for inactivated cells has been previously reported by Abdelmotilib et al. [59] and Bovo et al. [64], corroborating differences between viable and inactive cell adsorptions observed in most LAB strains. L. acidophilus and L. rhamnosus presented lower adsorption rates when compared to viable cells, probably due to the presence of cell walls containing a high number of denatured protein compounds and polysaccharides during thermal inactivation.The use of microbial adsorbents to complex AFM1 may provide an additional strategy to reduce its bioavailability [65]. Especially when evaluating the bioavailability of AFM1 in gastrointestinal models, where studies have demonstrated an increase in the binding between AFM1 and BAL through changes in the components of the cell envelope when exposed to bile [66,67,68]. Several strategies can be employed to improve the adsorption of aflatoxins by microorganisms, such as the use of inorganic based compounds. The use of inorganic adsorbents such as aluminosilicates, bentonites, and zeolites exhibit high in vitro aflatoxin adsorption capabilities when applied in dairy cattle [69,70]. The association of microorganisms with inorganic adsorbents has been the focus of several studies with the aim of mitigating the risk of mycotoxins in dairy cattle and concomitantly reducing AFM1 concentrations in milk. However, the use of adsorbents of mineral origin in raw milk may cause changes in the product's quality standards[69,71].The use of filtration steps can also contribute to reducing the bioavailability of AFM1 in milk, since the conjugate of BAL and AFM1 has the ability to be retained in filters with specific pore sizes. The filtration of the suspension of AFM1 and bacteria may increase the retention of AFM1 can also be enhanced as the conjugate accumulates in the filter due to the formation of a cake layer [72]. Despite favouring the removal of AFM1 from milk and ensuring safety for the consumer, filtration can also remove components from the product, damaging its nutritional quality and affecting the production of dairy products.None of the strains evaluated in the present study exhibited the potential for AFM1 biotransformation during the employed biotransformation assays under the established conditions. A re-extraction of the 2 μ L^{-1} concentration, in fact, confirmed that the toxin remained unchanged during the incubation process. However, although no biotransformation was observed, this effect cannot be neglected due to the enzymatic complex used for mycotoxin detoxification present in the employed LAB strains. Previous research has, in fact, documented the capability of various microorganisms to degrade mycotoxins. For example, depending on the degrading agent, aflatoxins can be subjected to several different degradation mechanisms including epoxidation, hydroxylation, dehydrogenation, and reduction [58,73,74,75].The adsorption observed herein demonstrates the potential application of these strains for AFM1 removal from milk, effectively reducing this compound's bioavailability. This demonstrates another use of LAB, aiming at contaminant removal alongside other technologies, enabling the incorporation of probiotic attributes into final products and preventing contaminant ingestion by consumers [29].

3.4.3 Antimicrobial bioremovals

Different antimicrobial concentrations were re-extracted, indicating that the antimicrobial remained unchanged during the incubation process. Thus, none of the evaluated strains presented antimicrobial bioremoval capacity under the established assay conditions. Despite no desired effect, several studies reporting antimicrobial bioremoval by LAB in different matrices are available, as reported by Lu et al [76].Research on the presence of antimicrobial residues in milk has grown exponentially over the years, denoting the increasing increase in the use of antimicrobial agents, as evidenced by Sachi et al. [77], demonstrating the growing concern regarding the risks associated with its presence in milk, which could generate several economic losses for the dairy sector and potential health risks for consumers. Several strategies have been developed with the aim of controlling and preventing the presence of antimicrobial residues in milk, such as: Prevention of discriminatory use of antimicrobials, development of highly sensitive detection tools, use of appropriate confirmatory methods for their quantification, development of public policies inspection systems capable of promptly identifying residues in milk, adequate inactivation of the antibiotics used capable of reducing the bioavailability of residues in the final product [77]. Among the forms of inactivation, the use of microorganisms to reduce bioavailability has been shown to be effective for removing various contaminants in the most diverse matrices, proving to be a potential method of mitigating antibiotic residues in milk [76].Although the thermal treatment used in the production of dairy derivatives is sufficient to reduce antimicrobial residue concentrations in final products, the results depend on several factors, such as the employed antimicrobial and applied temperature. The presence of residues in the final product and in raw milk can, therefore, pose risks to consumer health [42,78,79,80,81]. Future assessments are thus required to evaluate LAB as antimicrobial removal agents in milk.

4. Conclusion

The selected LAB assessed herein displays potential uses for many biotechnological applications in the dairy product industry. The probiotic characteristics of the investigated strains allow for their incorporation in various medical and food formulations, such as enteral feed, nutrient supplementation, and nutraceuticals. The observed lactose consumption rates indicate the possibility of obtaining low lactose content products while setting a precedent for studies focusing on the production of secondary metabolites of industrial interest, such as lactic acid. Despite no biometabolisation by the LAB strains investigated herein, the observed adsorption rates indicate the possibility of using these strains, especially L. acidophilus and L. casei, to recover contaminated milk, generating safe derivatives for human consumption. This, associated with antimicrobial resistance, suggests that the evaluated LAB can also still be assessed for the bioremoval of other antimicrobial contaminants. The findings indicate that all five strains investigated herein are promising biotechnological tools for the generation of low-lactose decontaminated milk derivatives.

Author Contributions: Conceptualisation, VFM, LAP, and FBNK; Methodology, VFM, LAP, FBNK, TRC, KMK, and LAMK; Validation, VFM, KMK, and LAMK; Formal analysis, VFM and LAMK; Investigation, VFM, LAP, FBNK and LAMK; Resources, KMK, ETM and LAMK; Data curation, VFM and LAP; Writing-original draft preparation, VFM, and LAMK; Writing-review and editing, LAP, FBNK, TRC, KMK, ETM and LAMK; Supervision, VFM, LAP and LAMK; Project administration, VFM; LAP, ETM and LAMK; Funding acquisition, ETM, LAMK. All authors have read and agreed to the published version of the manuscript.

Funding: This research was funded by FAPERJ - Fundação de Amparo à Pesquisa do Estado do Rio de Janeiro, Process E-16/010.002243/2019, and SEI 260003/002531/2021.

Acknowledgements: The authors would like to thank FAPERJ - Fundação de Amparo à

Pesquisa do Estado do Rio de Janeiro for scholarship and funding. Thanks are also due to the post-graduation course Veterinary Hygiene and Technology of Products of Animal Origin of the Federal Fluminense University. the Department of Food Technology at the same university (MTA-UFF), to the Mycology and Mycotoxin Laboratory at the Minas Gerais Federal University (LAMICO - UFMG).

Conflicts of Interest: The authors declare no conflict of interest.

References

1. Shehata, M. G.; El Sohaimy, S. A.; El-Sahn, M. A.; Youssef, M.M. Screening of isolated potential probiotic lactic acid bacteria for cholesterol lowering property and bile salt hydrolase activity. Ann. Agric. Sci. **2016,** 61, 65-75.
2. Byakika, S.; Mukisa, I.M.; Byaruhanga, Y.B.; Muyanja, C.A Review of Criteria and Methods for Evaluating the Probiotic Potential of Microorganisms. Food Rev. Int. **2019**. 35(5), 427 - 466.
3. Zmora, N.; Zilberman-Schapira, G.; Suez, J.; Mor, U.; Dori-Bachash, M.; Bashiardes, S.; Kotler, E.; Zur, M.; Regev-Lehavi, D.; Brik, R.B.; et al. Personalised gut mucosal colonization resistance to empiric probiotics is associated with unique host and microbiome features. Cell **2018**, 174, 388-1405.e2
4. Pinto, A.; Barbosa, J.; Albano, H.; Isidro, J.; Teixeira, P. Screening of Bacteriocinogenic Lactic Acid Bacteria and Their Characterisation as Potential Probiotics. Microorganisms **2020**, 8, 393.
5. Börner, R.A.; Kandasamy, V.; Axelsen, A.M.; Nielsen, A.T. Bosma, E.F. Genome editing of lactic acid bacteria: opportunities for food, feed, pharma and biotech. FEMS Microbio. Lett. **2019**. 366,fny291.
6. Cho, S.W.; Yim, J.; Seo, S.W. Engineering Tools for the Development of Recombinant Lactic Acid Bacteria.
Biotech. J. **2020.** 15, 1900344.
7. Wang, Y.; Wu, J.; Lv, M.; Shao, Z.; Hungwe, M.; Wang, J.; Bai, X.; Xie, J.; Wang, Y.; Geng, W. Metabolism Characteristics of Lactic Acid Bacteria and the Expanding Applications in Food Industry. Front. Bioeng. Biotechnol. **2021**, 9, 612285.
8. Jitpakdee, J.; Kantachote, D.; Kanzaki, H.; Nitoda, T. Potential of lactic acid bacteria to produce functional fermented whey beverage with putative health promoting attributes. LWT Food Sci. Technol. **2022**, 160, 113269.
9. Marcone, S.; Belton, O.; Fitzgerald, D.J. Milk-derived bioactive peptides and their health promoting effects: a potential role in atherosclerosis. Br J Clin Pharmacol. **2017,** 83(1), 152-162.
10. Rawoof, S. A. A.; Kumar, P. S.; Vo, D. V. N.; Devaraj, K.; Mani, Y.; Devaraj, T.; Subramanian, S. Production of optically pure lactic acid by microbial fermentation: a review. Envi. Chem. Lett. **2021**. 19, 539-556.
11. Hassan, N. S.; Jalil, A. A.; Hitam, C. N. C.; Vo, D. V. N.; Nabgan, W. Bio fuels and renewable chemicals production by catalytic pyrolysis of cellulose: a review. Environ. Chem. Lett. **2020**, 18, 1625 - 1648.
12. Srivastava, R. K.; Shetti, N. P.; Reddy, K. R.; Aminabhavi, T. M. Biofuels, biodiesel and biohydrogen production using bioprocesses: a review. Environ. Chem. Lett. **2020**,

18, 1049 - 1072.
13. McAuliffe, O. Symposium review: Lactococcus lactis from nondairy sources: their genetic and metabolic diversity and potential applications in cheese. J. Dairy Sci. **2018,** 101, 3597-3610.
14. Widyastuti, Y.; Febrisiantosa, A.; Tidona, F. Health-Promoting Properties of Lactobacilli in Fermented Dairy Products. Front. Microbiol. **2021**, 12, 673890.
15. Nishimura, J. Exopolysaccharides produced from Lactobacillus delbrueckii subsp. bulgaricus. Adv. Microbiol,
2014, 4, 1017-1023.
16. Daou, R.; Joubrane, K.; Maroun, R.G.; Khabbaz, L.R.; Ismail, A.; El Khoury, A. Mycotoxins: Factors influencing production and control strategies AIMS Agricult. Food **2021**, 6(1), 416 - 447.
17. Ismail, A.; Gonçalves, B. L.; de Neeff, D.V.; Ponzilacqua, B.; Coppa, Carolina F.S.C.; Hintzsche, H.; Sajid, M.; Cruz, A.G.; Corassin, C.H.; Oliveira, C.A.F.Aflatoxin in foodstuffs: Occurrence and recent advances in decontamination. Food Res. Intern. **2018.** 113, 74-85.
18. Sadiq, F.A.; Yan, B.; Tian, F.; Zhao, J.; Zhang, H.; Chen, W. Lactic Acid Bacteria as Antifungal and Anti- Mycotoxigenic Agents: A Comprehensive Review Rev. Food Sci. Food Saf. **2019**, 18(5), 1403 - 1436.
19. Petrova, P.; Arsov, A.; Tsvetanova, F.; Parvanova-ancheva, T.; Vasileva, E.; Tsigoriyna, L.; Petrov K. The Complex Role of Lactic Acid Bacteria in Food Detoxification. Nutrients, **2022**, 14(10), 2038.
20. Saeki, E.K.; Farhat, L.P.; Pontes, E.A. Efficiency of glycerol and skimmed milk cryoprotectants for freezing microorganisms. Acta Vet. Br. **2015.** 9(2), 195 - 198.
21. Malik, M.; Bora, J.; Sharma, V. Growth studies of potentially probiotic lactic acid bacteria (Lactobacillus plantarum, Lactobacillus acidophilus, and Lactobacillus casei) in carrot and beetroot juice substrates. J. Food process preserv. **2019.** 00, e14214.
22. APHA. American Public Health Association. Compendium of Methods for the Microbiological Examination of foods. 3 ed. Washington. **2015**. p. 1219.
23. Zhang, G.; Li, J.; Lv, J.; Liu, L.; Li, C.; Liu, L. Decontamination of aflatoxin M1 in yogurt using Lactobacillus rhamnosus LC-4. J Food Saf. **2019**; 39, e12673.
24. Jensen, H.; Grimmer, S.; Naterstad, K.; Axelsson, L. In vitro testing of commercial and potential probiotic lactic acid bacteria. Int. J. Food Microbiol. **2012,** 153, 216 - 222.
25. CLSI - Clinical and Laboratory Standards Institute. M07-A9: Methods for Dilution Antimicrobial Susceptibility Tests for Bacteria That Grow Aerobically; Approved Standard-Ninth Edition, v. 32, n. 2. Pennsylvania: CLSI Wayne, **2012**. 88p.
26. Brazil. Ministry of Health. National Health Surveillance Agency (ANVISA). Resolution of the Collegiate Board of Directors - RDC no. 253, of 16 September 2003. Creates the Programme for the Analysis of Residues of Veterinary Medicines in Foodstuffs of Animal Origin - PAMVet. Diário Oficial [da] União, Brasília, DF, n. 181, p. 90, 18 Sep. **2003**, a. Section 1.
27. Ni, Y.; Huang, C.; Kokot, S. A kinetic spectrophotometric method for the determination of ternary mixtures of reducing sugars with the aid of artificial neural networks and multivariate calibration. Anal. Chim. Acta. **2003,** 480, 53 - 65.
28. Shundo, L.; Ruvieri, V.; Navas, S. A.; Sabino, M. Optimisation of aflatoxin M1

determination in milk using immunoaffinity column and thin layer chromatography. Rev. Inst. Adolfo Lutz. **2004.** 63(1) 43 - 48.
29. Keller, L.A.M.; Abrunhosa, L.; Keller, K.M.; Rosa, C.A.R.; Cavaglieri, L.R.; Venâncio, A. ZENralenone and its derivatives α-ZENralenol and β-ZENralenol decontamination by Saccharomyces cerevisiae strains isolated from bovine forage. Toxins. **2015.** 7(8), 3297 - 3308.
30. Moebus, V.F.; Pinto, L.d.A.; Köptcke, F.B.N.; Keller, K.M.; Aronovich, M.; Keller, L.A.M. In Vitro Mycotoxin Decontamination by Saccharomyces cerevisiae Strains Isolated from Bovine Forage. Fermentation **2023**, 9, 585.
31. Massoud, R.; Darani, K. K.; Sharifan, A.; Asadi, G. H.; Hadiani, M. R. Mercury Biodecontamination from Milk by using L. acidophilus ATCC 4356. J. Pure Appl. Microbiol. **2020**, 14(4), 2313 - 1321.
32. Brazil. Ministry of Health. National Health Surveillance Agency (ANVISA). Brazilian Pharmacopoeia, 6 ed. volume 2. **2019**. p. 1512.
33. FAO/WHO Working Group. Report on Drafting Guidelines for the Evaluation of Probiotics in Food: Guidelines for the Evaluation of Probiotics in Food. London. **2002,** p. 11.
34. Brazil. Resolution RDC no. 2, of 07 January 2002. Approves the Technical Regulation on Bioactive Substances and Isolated Probiotics with Claims of Functional and/or Health Properties. D.O.U. - Diário Oficial da União, Brasília, DF, 09 January **2002.**
35. Grumet, L.; Tromp, Y.; Stiegelbauer, V. The Development of High-Quality Multispecies Probiotic Formulations: From Bench to Market. Nutrient. **2020**, 12, 2453
36. Muaz, K.; Riaz, M. Decontamination of aflatoxin m1 in milk through integration of microbial cells with sorbitan monostearate, activated carbon and bentonite. J. Anim. Plant Sci. **2021,** 31(1), Page: 235-245
37. Luo, Y., Liu, X., Yuan, L.; Li, J. Complicated interactions between bio-adsorbents and mycotoxins during mycotoxin adsorption: Current research and future prospects. Trends Food Sci Technol **2020**, 96, 127 - 134.
38. Muaz, K.; Riaz, M.; Rosim, R.E.; Akhtar, S.; Corassin, C.H.; Gonçalves, B.L; Oliveira, C.A.F. In vitro ability of nonviable cells of lactic acid bacteria strains in combination with sorbitan monostearate to bind to aflatoxin M1 in skimmed milk. LWT Food Sci. Technol**. 2021.** 147, 111666.
39. Ayivi, R.D.; Gyawali, R.; Krastanov, A.; Aljaloud, S.O.; Worku, M.; Tahergorabi, R.; Silva, R.C.d.; Ibrahim, S.A. Lactic Acid Bacteria: Food Safety and Human Health Applications. Dairy **2020**, 1, 202-232.
40. Wieërs, G.; Belkhir, L.; Enaud, R.; Leclercq, S.; de Foy, J.M.P.; Dequenne, I.; de Timary, P.; Cani, P.D. How Probiotics Affect the Microbiota. Front. Cell. Infect. Microbiol. **2020.** 9, 454.
41. Treiber, F.M.; Beranek-Knauer, H. Antimicrobial Residues in Food from Animal Origin-A Review of the Literature Focusing on Products Collected in Stores and Markets Worldwide. Antibiotics **2021**, 10, 534.
42. Virto, M.; Santamarina-García, G.; Amores, G.; Hernández, I. Antibiotics in Dairy Production: Where Is the Problem? Dairy **2022**, 3, 541-564.
43. Chiesa, L.M.; DeCastelli, L.; Nobile, M.; Martucci, F.; Mosconi, G.; Fontana, M.; Castrica, M.; Arioli, F.; Panseri,S. Analysis of antibiotic residues in raw bovine milk

and their impact toward food safety and on milk starter cultures in cheese-making process. LWT Food Sci. Technol. **2020**, 131, 109783.

44. Bacanlı, M.; Başaran, N. Importance of antibiotic residues in animal food. Food Chem. Toxicol. **2019**, 125, 462- 466.
45. Devika, J.D.; Aparna, S.; Johnson, J.B.; Sabu, T. Critical insights into antibiotic resistance transferability in probiotic Lactobacillus. Nutrition **2020**, 69, 110567
46. Li, T.; Teng, D.; Mao, R.; Hao, Y.; Wang, X.; Wang, J. A critical review of antibiotic resistance in probiotic bacteria. Food Res. Int. **2020**, 136 109571
47. Bintsis, T. Lactic acid bacteria: their applications in foods. J Bacteriol Mycol. **2018**, 6(2), 89 - 94.
48. Siroli1, L.; Patrignani, F.; Serrazanetti, D.I.; Parolin, C.; Palomino, R.A.Ñ.; Vitali, B.; Lanciotti, R. Determination of Antibacterial and Technological Properties of Vaginal Lactobacilli for Their Potential Application in Dairy Products. Front. Microbiol**. 2018.** 8, 1.
49. Savaiano, D.A.; Hutkins, R.W. Yoghurt, cultured fermented milk, and health: A systematic review. Nutrition reviews **2021.** 79(5), 599-614
50. Schulz, P.; Rizvi, S. S. H. Hydrolysis of Lactose in Milk: Current Status a Future Products. Food Rev. Int. **2021.**
51. Timothy, B.; Iliyasu, A.H.; Anvikar, A.R. Bacteriocins of Lactic Acid Bacteria and Their Industrial Application.
Curr. Top. Lact. Acid Bact. Probiotics **2021**,7(1), 1-13.
52. Barbieri, F.; Montanari, C.; Gardini, F.; Tabanelli, G. Biogenic Amine Production by Lactic Acid Bacteria: A Review. Foods **2019**, 8, 17.
53. Shao, X.; Xu, B.; Chen, C.; Li, P.; Luo, H. The function and mechanism of lactic acid bacteria in the reduction of toxic substances in food: a review, Crit. Rev. Food. Sci. Nutr. **2022** 62(21), 5950-5963.
54. Bartkiene, E.; Jakobsone, I.; Juodeikiene, G.; Vidmantiene, D.; Pugajeva, I.; Bartkevics, V. Effect of fermented Helianthus tuberosus L. tubers on acrylamide formation and quality properties of wheat bread. LWT - Food Sci. and Tech. **2013a**, 54(2), 414-20.
55. Bartkiene, E.; Jakobsone, I.; Juodeikiene, G.; Vidmantiene, D.; Pugajeva, I.; Bartkevics, V. Study on the reduction of acrylamide in mixed rye bread by fermentation with bacteriocin-like inhibitory substances producing lactic acid bacteria in combination with Aspergillus niger glucoamylase. Food Control **2013b** 30(1), 35 - 40.
56. IARC Some naturally occurring substances: Food items and constituents, heterocyclic aromatic amines and mycotoxins IARC Monographs on the Evaluation of Carcinogenic Risks to Humans **1993**, 56, 1-599.
57. Turna, N.S.; Wu, F. Aflatoxin M1 in milk: A global occurrence, intake, & exposure assessment. Trends Food Sci. Technol**. 2021.** 110, 183-192.
58. Li, P.; Su, R.; Yin, R.; Lai, D.; Wang, M.; Liu, Y.; Zhou, L. Detoxification of Mycotoxins through Biotransformation. Toxins **2020**, 12, 121.
59. Abdelmotilib, N.M.; Hamad, G.; Salem, E.G. Aflatoxin M1 reduction in milk by a novel combination of probiotic bacterial and yeast strains. Nutr. Food Saf. **2018**. 8, 83-99.
60. Assaf, J.C.; Khoury, A.E.; Chokr, A.; Louka, N.; Atoui, A. A novel method for

elimination of aflatoxin M1 in milk using Lactobacillus rhamnosus GG biofilm. Intern. J. Dairy Tech. **2019**. 72, 248-256.
61. Martínez, M.; Magnoli, A.; Pereyra, M.G.; Cavaglieri, L. Probiotic bacteria and yeasts adsorb aflatoxin M1 in milk and degrade it to less toxic AFM1-metabolites. Toxicon **2019**. 172, 1-7.
62. Pfliegler, W.P., Pusztahelyi, T.; Pócsi, I. Mycotoxins-prevention and decontamination by yeasts. J. Basic Microbiol. **2015**, 55, 805 - 818.
63. Luo, Y., Liu, X., Yuan, L.; Li, J. Complicated interactions between bio-adsorbents and mycotoxins during mycotoxin adsorption: Current research and future prospects. Trends Food Sci Technol **2020**, 96, 127 - 134.
64. Bovo, F., Corassin, C.H., Rosim, R.E.; Oliveira, C.A.F. Efficiency of Lactic Acid Bacteria Strains for Decontamination of Aflatoxin M1 in Phosphate Buffered Saline Solution and in Skimmed Milk. Food Biopr. Tech. **2013**, 6, 2230-2234.
65. Ismail, A.; Akhtar, S.; Levin, R.E.; Ismail, T.; Riaz, M.; Amir, M. Aflatoxin M1: Prevalence and decontamination strategies in milk and milk products. Crit. Rev. Microbiol. **2015**, 42, 418-427.
66. Peltonen, K.D.; El-Nezami, H.; Haskard, C.; Ahokas, J.; Salminen, S. Aflatoxin B1 Binding by Dairy Strains of Lactic Acid Bacteria and Bifidobacteria. Dairy Sci. **2001**, 84, 2152-2156.
67. El-Nezami, H.; Kankaanpää, P.; Salminen, A.J. Physicochemical alterations enhance the ability of dairy strains of lactic acid bacteria to remove aflatoxin from contaminated media. Food Prot. **1998**, 61, 466-468.
68. Zhao, L.; Wei, J.; Zhao, H.; Zhu, B.; Zhang, B. Detoxification of carcinogenic compounds by lactic acid bacteria strains. Crit. Rev. Food Sci. Nutr. **2018**, 58, 2727-2742.
69. Vila-Donat, P., Marín, S.; Ramos, A.J. A review of the mycotoxin adsorbing agents, with an emphasis on their multi-binding capacity, for animal feed decontamination. Food Chem. Toxicol. **2018**, 114, 246 - 259.
70. Ahn, J.Y., Kim, J., Cheong, D.H., Hong, H., Jeong, J.Y.; Kim, B.G. An In Vitro Study on the Efficacy of Mycotoxin Sequestering Agents for Aflatoxin B1, Deoxynivalenol, and Zearalenone. Animals **2022**, 12, 333.
71. Xiong, J.L., Wang, Y.M., Zhou, H.L.; Liu, J.X. Effects of dietary adsorbent on milk aflatoxin M1 content and the health of lactating dairy cows exposed to long-term aflatoxin B1 challenge. J. Dairy Sci. **2018**, 101, 8944 - 8953.
72. Assaf, J.C.; Atoui, A.; El Khoury, A.; Chokr, A.; Louka, N. A comparative study of procedures for binding of aflatoxin M1 to Lactobacillus rhamnosus GG. Brazilian J. Microbiol. **2017**, 49, 120-127.
73. Gao, X.; Ma, Q.; Zhao, L.; Lei, Y.; Shan, Y.; Ji, C. Isolation of Bacillus subtilis: screening for aflatoxins B1, M1, and G1 detoxification Eur. Food Res. Technol. **2011**, 232(6), p. 957.
74. Zhao, L.H.; Guan, S.; Gao, X.; Ma, Q.G.; Lei, Y.P.; Bai, X.M.; Ji, C. Preparation, purification and characteristics of an aflatoxin degradation enzyme from Myxococcus fulvus ANSM068. J. Appl. Microbiol. **2011.** 110 (1), 147- 155.
75. Luo, Y.; Wang, J.G.; Liu, B.; Wang, Z.L.; Yuan, Y.H.; Yue, T.L. Effect of yeast cell morphology, cell wall physical structure and chemical composition on patulin adsorption. PLoS ONE **2015**, 10, e0136045.

76. Lu, Z.Y.; Ma, Y.L.; Zhang, J.T.; Fan, N.S.; Huang, B. C.; Jin, R.C.; A critical review of antibiotic removal strategies: Performance and mechanisms, J. Water Process Eng. **2020.** 38,101681.
77. Sachi, S.; Ferdous, J.; Sikder, M. H.; Hussani, S. M. A. K. Antibiotic residues in milk: Past, present, and future.
J. Adv. Vet. Anim. Res. **2019**, 6(3), 315-332.
78. Hassan, H. F.; Saidy, L.; Haddad, R.; Hosri, C.; Asmar, S.; Jammoul, A.; Jammoul, R.; Hassan, H.; Serhan, M. Investigation of the effects of some processing conditions on the fate of oxytetracycline and tylosin antibiotics in the making of commonly consumed cheeses from the East Mediterranean. Vet. World **2021**, 14, 1644 - 1649.
79. Tian, L.; Khalil, S.; Bayen, S. Effect of thermal treatments on the degradation of antibiotic residues in food.
Crit. Rev. Food Sci. Nutr. **2017**, 57, 3760 - 3770.
80. Quintanilla, P.; Beltrán, M. C.; Molina, A.; Escriche, I.; Molina, M. P. Characteristics of ripened Tronchón cheese from raw goat milk containing legally admissible amounts of antibiotics. J. Dairy Sci. **2019**, 102, 2941 - 2953.
81. Gajda, A.; Nowacka-Kozak, E.; Gbylik-Sikorska, M.; Posyniak, A. Tetracycline antibiotics transfer from contaminated milk to dairy products and the effect of the skimming step and pasteurisation process on residue concentrations. Food Addit. Contam. Part A **2018**, 35, 66 - 76.

6. CONCLUSIONS AND FINAL CONSIDERATIONS

6.1 CONCLUSION OF DEVELOPMENT 1

The isolation of wild microorganism strains is a promising alternative for the development of new anti-mycotoxin additives with probiotic functions, reinforcing the idea of using additives of biological origin. The S. cerevisiae strains evaluated showed excellent adsorption capacity for AFB1 and ZEN under gastrointestinal conditions.

All the strains showed high potential for commercial applications, especially strain LL08, but further studies are needed to optimise it for use in livestock farming. Although no significant biotransformation was observed, this effect should not be ruled out when selecting microorganisms with anti-mycotoxin potential, as many microorganisms can convert mycotoxins into less toxic metabolites in the long term, which is important for the development of new products.

As microorganisms have a lower adsorptive capacity compared to inorganic adsorbents, commercial products must be developed with a focus on the genomic selection of biotransformation enzymes aimed at detoxifying mycotoxins. The elimination of mycotoxins using additives is not, however, sufficient to guarantee the safety of animal feed, and mechanisms to control and select quality raw materials must also be applied to improve product quality and animal health.

6.2 CONCLUSION OF DEVELOPMENT 2

The selected LAB have the potential for various biotechnological applications in the dairy industry. The probiotic characteristics of the strains investigated allow them to be incorporated into various medical and food formulations, such as enteral feeds, nutrient supplements and nutraceuticals. The lactose consumption rates observed indicate the possibility of obtaining products with a low lactose content and set a precedent for studies focusing on the production of secondary metabolites of industrial interest, such as lactic acid.

Although there was no biometabolisation by the LAB strains investigated here, the adsorption rates observed indicate the possibility of using these strains, especially L. acidophilus and L. casei, to recover contaminated milk, generating derivatives that are safe for human consumption.human consumption. This, coupled with antimicrobial resistance, suggests that the LAB evaluated can also be assessed for the bioremoval of other antimicrobial contaminants. The results indicate that all five strains investigated are promising biotechnological tools for generating decontaminated dairy products with low lactose content.

6.3 FINAL CONSIDERATIONS

The overall work involved a huge amount of data, resulting in the two articles presented in this thesis and other data that is in the final stages of being tabulated and statistically analysed. Once finalised, it is planned to write and submit at least two articles from this thesis.
Eliminating mycotoxins through additives is not enough to guarantee food safety, and control mechanisms and the selection of quality raw materials must also be applied to improve product quality and animal health.
The S. cerevisiae strains evaluated showed excellent adsorption capacity for AFB1 and ZEN under gastrointestinal conditions when compared to the commercial product.
The isolation of microorganisms from nature is a promising alternative for the development of new antimycotoxin additives with probiotic functions, reinforcing the idea of using additives of biological origin.
Since microorganisms have a lower adsorptive capacity for AFB1 compared to inorganic adsorbents, commercial products should be developed with a focus on the genomic selection of biotransformation enzymes for mycotoxin detoxification.
Although no significant biotransformation was observed, this effect should not be ruled out when selecting microorganisms with antimycotoxin potential, as many microorganisms can convert mycotoxins into less toxic metabolites in the long term, which is important for the development of new products.
The use of fermenting microorganisms has proved to be a viable and sustainable alternative for removing contaminants from various matrices

Although biotransformation was not observed, the search for microorganisms with the enzymatic machinery is relevant. The methodology implemented in this work will serve to evaluate the potential applicability of new strains
The strains used showed excellent adsorption capacity and could serve as a basis for developing new products
Lactose fermentation expands the universe of biotechnological applications for LAB and can be used in various areas.
The probiotic characteristics of the strains allow them to be incorporated into various medical and food formulations to serve consumers in different segments, such as enteral feeding, nutritional supplementation and nutraceuticals.
The development of new techniques in order to improve the selection of new microorganisms must be carried out continuously in order to obtain a product with potential commercial applicability.
The selection of probiotic strains of Lactobacillus spp. opens up the possibility of obtaining new products with a wide range of applicability, whether as an anti-mycotoxin additive or with industrial applications to obtain secondary metabolites of interest, such as lactic acid and bacteriocins.
Future research will be carried out during the post-doctoral period focussing on the identification, quantification and isolation of secondary metabolites from BAL with subsequent industrial application in the food and drug industries.

7 BIBLIOGRAPHICAL REFERENCES

AAZAMI, M. H.; NASRI, M. H. F.; MOJTAHEDI, M.; MOHAMMADI, S. R. In vitro aflatoxin B1 binding by the cell wall and (1→3)-β-d-glucan of baker's yeast, Journal of food protection, n.81, p. 670-676. 2018.

ABDEL-KAREEM, M. M.; RASMEY, A. M.; ZOHRI, A. A. The action mechanism and biocontrol potentiality of novel isolates of Saccharomyces cerevisiae against the aflatoxigenic Aspergillus flavus. Letters of Applied Microbiology, n. 68, p. 104 - 111. 2019.

ABDELMOTILIB, N. M.; HAMAD, G.; SALEM, E. G. Aflatoxin M1 reduction in milk by a novel combination of probiotic bacterial and yeast strains. Nutrition and Food Safety, v. 8, p. 83-99. 2018.

ADEGBEYE, M. J.; REDDY, P. R. K.; CHILAKA; C. A.; BALOGUN, O. B.; ELGHANDOUR, M. M. M. Y.; RIVAS-CACERES, R. R.; SALEM, A. Z. M. Mycotoxin toxicity and residue in animal products: Prevalence, consumer exposure and reduction strategies - A review. Toxicon, v. 177, p. 96 - 108. 2020

ADEYEYE, S. A. O. Aflatoxigenic fungi and mycotoxins in food: a review. Critical reviews in food science and nutrition, p. 1-13. 2019.

AHN, J. Y.; KIM, J.; CHEONG, D. H.; HONG, H.; JEONG, J. Y.; KIM, B. G. An In Vitro Study on the Efficacy of Mycotoxin Sequestering Agents for Aflatoxin B1, Deoxynivalenol, and Zearalenone. Animals, v. 12, p. 333. 2022.

ALAM S.: DENG Y. Protein interference on aflatoxin B1 adsorption by smectites in corn fermentation solution. Applied Clay Science, n. 144, p. 36 - 44. 2017.

ALFANDARI, S.; CANNESSON, O. Aminoglucosides. EMC - Treatise on Medicine, v. 20, n. 3, p. 1-4. 2016.

ALMEIDA NETO, J. R. M.; SANTOS, G. M.; ARROYO, R. J. O.; SOUSA, V. O.; FERREIRA, A. M. Sustainability of small dairy farms. Revista Interdisciplinar de Direito, v. 10, n. 2, p. 397 - 402. 2017.

ALVITO, P. C.; SIZOO, E. A.; ALMEIDA, C. M. M.; van EGMOND, H. P. Occurrence of Aflatoxins and Ochratoxin A in Baby Foods in Portugal. Food Analitycal Methods, v. 3, p. 22 - 30. 2010.

ANDO, H.; HATANAKA, K.; OHATA, I.; YAMASHITA-KITAGUCHI, Y.; KURATA, A.; KISHIMOTO, N. Antifungal activities of volatile substances generated by yeast isolated from Iranian commercial cheese. Food Control, n. 26, p. 472-478. 2012.

ANFOSSI, L.; GIOVANNOLI, C.; BAGGIANI, C. Mycotoxin detection. Current Opinion in Biotechnology, v. 37, p. 120-126. 2016.

APHA. AMERICAN PUBLIC HEALTH ASSOCIATION. Compendium of Methods for the Microbiological Examination of foods. 3 ed. Washington. 2015, 1219p.

ARMANDO, M. R.; PIZZOLITTO, R. P.; DOGI, C. A.; CRISTOFOLINI, A.; MERKIS, C.; POLONI, V.; DALCERO, A. M.; CAVAGLIERI, L. R. Adsorption of ochratoxin A and zearalenone by potential probiotic Saccharomyces cerevisiae strains and its relation with cell wall thickness. Journal of Applied Microbiology, v. 113, p. 256-264. 2012.

ASSAF, J. C.; ATOUI, A.; KHOURY, A. E.; CHOKR, A.; LOUKA, N. A comparative study of procedures for binding of aflatoxin M1 to Lactobacillus rhamnosus GG. Brazilian Journal of Microbiology, v. 49, p. 120-127. 2017.

ASSAF, J. C.; KHOURY, A. E.; CHOKR, A.; LOUKA, N.; ATOUI, A. A novel method for elimination of aflatoxin M1 in milk using Lactobacillus rhamnosus GG biofilm. International Journal of Dairy Technology, v. 72, p. 248-256. 2019.
AYIVI, R. D.; GYAWALI, R.; KRASTANOV, A.; ALJALOUD, S. O.; WORKU, M.; TAHERGORABI, R.; SILVA, R. C. D.; IBRAHIM, S. A. Lactic Acid Bacteria: Food Safety and Human Health Applications. Dairy, v. 1, p. 202-232. 2020.
AYTENFSU, S.; MAMO, G.; KEBEDEL, B. Review on Chemical Residues in Milk and Their Public Health Concern in Ethiopia. Journal of Nutrition & Food Sciences, v. 6, n 4. 2016.Bacanlı, M.; Başaran, N. Importance of antibiotic residues in animal food. Food Chem. Toxicol. 2019, 125, 462-466.
BAHAR, A. A.; REN, D. Antimicrobial peptides. Pharmaceuticals, n. 6, p. 1543-1575. 2013.
BANDEIRA, M. G. L.; SANTOS, A. S.; ABRANTES, M. R.; REBOUÇAS, G. G.; SILVA, M. E. T.; PAIVA, W. S.; MAIA, M. O.; LIMA, L. S. C.; SILVA, J. B. A.; DAMACENO, M. N. Sensitivity profile of Staphylococcus spp. isolated from food to pharmaceutical antibiotics. In: Proceedings of the XII Latin American Congress on Food Microbiology and Hygiene. Blucher Food Science Proceedings, v. 1, n. 1, p. 23-24, São Paulo, 2014.
BANDO, E.; OLIVEIRA, R. C.; FERREIRA, G. M.; MACHINSKI, M. Occurrence of antimicrobial residues in pasteurised milk commercialized in the state of Parana, Brazil. Journal of Food Protection, v. 72, p. 911-914, 2009.
BANGAR, S. P.; SHARMA, N.; KUMAR, M.; OZOGUL, F.; PUREWAL, S. S.; TRIF, M. Recent developments in applications of lactic acid bacteria against mycotoxin production and fungal contamination. Food Bioscience, n. 44, p. 101444. 2021.
BARBIERI, F.; MONTANARI, C.; GARDINI, F.; TABANELLI, G. Biogenic Amine Production by Lactic Acid Bacteria: A Review. Foods, v. 8, p. 17. 2019.
BARRIENTOS-VELÁZQUEZ A. L.; ARTEAGA S.; DIXON J. B.; DENG Y. The effects of pH, pepsin, cation exchange, and vitamins on aflatoxin adsorption on smectite in simulated gastric fluids. Applied Clay Science, n. 120, p. 17-23. 2016.
BARTKIENE, E.; JAKOBSONE, I.; JUODEIKIENE, G.; VIDMANTIENE, D.; PUGAJEVA,I.; BARTKEVICS, V. Effect of fermented Helianthus tuberosus L. tubers on acrylamide formation and quality properties of wheat bread. LWT - Food Science and Technology, v. 54,n. 2, p. 414-20. 2013a.

BARTKIENE, E.; JAKOBSONE, I.; JUODEIKIENE, G.; VIDMANTIENE, D.; PUGAJEVA,I.; BARTKEVICS, V. Study on the reduction of acrylamide in mixed rye bread by fermentation with bacteriocin-like inhibitory substances producing lactic acid bacteria in combination with Aspergillus niger glucoamylase. Food Control, v. 30, n. 1, p. 35 - 40. 2013b.
BATISTA, R. A. B; ASSUNÇÃO, D. C. B.; PENAFORTE, F. R. O.; JAPUR, C. C. Lactose in industrialised foods: evaluation of the availability of quantity information. Ciência & Saúde Coletiva, v. 23, n. 12, p. 4119-4128. 2018
BEGOT, C.; DESNIER, I.; DAUDIN, J. D.; LABADIE, J. C.; LEBERT, A. Recommendations for calculating growth parameters by optical density measurements. Journal of Microbiological Methods, v.25, n.3, p.225-232, 1996
BELOGLAZOVA, N.V.; EREMIN, S.A. Rapid screening of aflatoxin B1 in beer by fluorescence polarisation immunoassay. Talanta, v.142, p. 170-175. 2015.

BENKERROUM, N. Chronic and acute toxicities of aflatoxins: mechanisms of action. International Journal of Environmental Research and Public Health, v. 17, n. 423, p. 1-28. 2020.

BERENDSEN, B.; STOLKER, L.; DE JONG, J.; NIELEN, M.; TSERENDORJ, E.; SODNOMDARJAA, R.; CANNAVAN, A.; ELLIOTT, C. Evidence of natural occurrence of the banned antibiotic chloramphenicol in herbs and grass. Analytical and Bioanalytical Chemistry, v. 397, n. 5, p. 1955-1963, 2010.

BINTSIS, T. Lactic acid bacteria: their applications in foods. Journal of Bacteriology and Mycology, v. 6, n. 2, p. 89 - 94. 2018.

BLAIR, J. M. A.; WEBBER, M. A.; BAYLAY, A. J.; OGBOLU, D. O.; PIDDOCK, L. J. V. Molecular mechanisms of antibiotic resistance. Nature reviews Microbiology, n. 13, p. 42-51. 2015.

BORDALO TONUCCI, L.; DOS SANTOS, K. M. O.; FERREIRA, C. L. L. F. F.; RIBEIRO, S. M. R.; OLIVEIRA, L. L.; MARTINO, H. S. D. Gut microbiota and probiotics: focus on diabetes mellitus. Critical Reviews in Food Science and Nutrition, v. 57, n. 11, p. 2296 - 2309. 2017.

BÖRNER, R. A.; KANDASAMY, V.; AXELSEN, A. M.; NIELSEN, A. T.; BOSMA, E. F. Genome editing of lactic acid bacteria: opportunities for food, feed, pharma and biotech. FEMS Microbiology Letters, v. 366, p. fny291. 2019.

BOUDERGUE, C.; BUREL, C.; DRAGACCI, S.; FRAVOT, M. C.; FREMY, J. M.; MASSIMI, C.; PRIGENT, P.; DEBONGNIE, P.; PUSSEMIER, L.; BOUDRA, H.; MORGAVI, D. P.; OSWALD, I. P.; PEREZ, A. AVANTAGGIATO, G. Review of mycotoxin-detoxifying agents used as feed additives: mode of action, efficacy and feed/food safety. EFSA Supporting Publications, v.6, n.9. 2009.

BOVO, F.; CORASSIN, C. H.; ROSIM, R. E.; OLIVEIRA, C. A. F. Efficiency of lactic acid bacteria strains for decontamination of alfatoxin M1 in phosphate buffer saline solution. Food Bioprocess Technology, v. 6, p. 2230 - 2234. 2013.

BRAZIL. Ministry of Agriculture and Supply. Normative Instruction no. 42, of 20 December 1999. Amends the National Plan for the Control of Residues in Products of Animal Origin - PNCR and the Programmes for the Control of Residues in Meat - PCRC, Honey - PCRM, Milk - PCRL and Fish - PCRP. Diário Oficial [da] União, Brasília, DF, n. 181, p. 253, 22 Dec. 1999. Section 1.

BRAZIL. MINISTRY OF HEALTH. NATIONAL SURVEILLANCE AGENCY SANITARY (ANVISA). Resolution RDC no. 2, of 7th January 2002. Approves the technical regulation on bioactive substances and isolated probiotics with functional and/or health claims. Diário Oficial [da] República Federativa do Brasil, Brasília, DF, n. 136, p. 78, 09 Jan. 2002. Section 1.

BRAZIL. Ministry of Health. National Health Surveillance Agency (ANVISA). Collegiate Board Resolution - RDC no. 253, of 16 September 2003. Creates the Programme for the Analysis of Residues of Veterinary Medicines in Foodstuffs of Animal Origin - PAMVet. Diário Oficial [da] União, Brasília, DF, n. 181, p. 90, 18 Sep. 2003a. Section 1.

BRAZIL. Ministry of Agriculture, Livestock and Supply. Ordinance no. 130, of 24 May 2006. Setting up the Working Group on Mycotoxins in products intended for animal feed. Official Gazette, Brasília, DF, p. 5, 25 May 2006. Section 2.

BRAZIL. Ministry of Health. Anvisa. National Health Surveillance Agency (ANVISA).

Programme for the Analysis of Residues of Veterinary Medicines in Foodstuffs (PAMvet). Report 2006-2007, 2009.
BRAZIL. MINISTRY OF HEALTH. National Food and Nutrition Policy. Brasilia. 2013.
BRAZIL. MINISTRY OF HEALTH. Food guide for the Brazilian population: promoting healthy eating. Brasília. 2014.
BRAZIL. Ministry of Agriculture, Livestock and Supply. Manual of Official Methods for Analysing Food of Animal Origin. Brasilia. 2018.
BRAZIL. Ministry of Agriculture, Livestock and Supply. Normative Instruction - IN no. 76, of 26 November 2018. The Technical Regulations that establish the identity and quality characteristics of refrigerated raw milk, pasteurised milk and type A pasteurised milk are approved. Diário Oficial [da] União, Brasília, DF, n. 230, p. 9, 30 Nov. 2018. Section 1.
BRAZIL. Ministry of Health. National Health Surveillance Agency (ANVISA). Farmacopeia Brasileira, 6 ed. volume 2. 2019. 1512p.
BRAZIL. Ministry of Health. National Health Surveillance Agency (ANVISA). Resolution of the collegiate board - RDC no. 60, of 23 December 2019. Establishes the lists of microbiological standards for foods. Diário Oficial da União, Brasília, DF, n. 249, p. 133, 26 Dec. 2019. Section 1.

BRAZIL. Presidency of the Republic, General Secretariat, Sub-Cabinet for Legal Affairs. health. Decree no. 10.468, of 18 August 2020. Amends Decree No. 9.013, of 29 March 2017, which regulates Law No. 1.283, of 18 December 1950, and Law No. 7.889, of 23 November 1989, which provide for the regulation of the industrial and sanitary inspection of products of animal origin. Diário Oficial [da] União, Brasília, DF, n. 159, p. 5, 19 Aug. 2020. Section 1.
BRAZIL, Embrapa. Milk Yearbook 2021. [S.l: s.n.], 2021. 116 p. Available at: <www.embrapa.br/gado-de-leite>. Accessed 21 Dec. 2022.
BRAZIL. Ministry of Health. National Health Surveillance Agency. NORMATIVE INSTRUCTION - IN NO. 160, OF 1ST JULY 2022. Establishes the maximum tolerated limits (MRL) for contaminants in food. Official Gazette, Brasília, DF, n. 126, 06 July 2022. Section 1.
BRAZIL. BRAZILIAN INSTITUTE OF GEOGRAPHY AND STATISTICS - IBGE. Survey Quarterly Milk Milk - 1º quarter 2023. Availableat: <https://sidra.ibge.gov.br/home/leite/rio-de-janeiro>. Accessed 21 Dec. 2022.
BRETAS, A. de A. Inclusion of mycotoxin adsorbents for piglets. CES Medicina veterinaria y zootecnia, v. 13, n. 1, p. 80-95. 2018.
BRUNTON, L. L.; CHABNER, B. A.; KNOLLMAN, B. C. As Bases Farmacológicas da Terapêutica de Goodman & Gilman 12 ed. São Paulo: Mc Graw Hill Brasil. 2012. 2112 p.
BUSTAMANTE, M.; OOMAH, B. D.; OLIVEIRA, W. P.; BURGOS-DÍAZ, C.; RUBILLAR,
M.; SHENE, C. Probiotics and prebiotics potential for the care of skin, female urogenital tract, and respiratory tract. Folia Microbiológica, v. 65, n. 2, p. 245 - 264. 2020.
BUZALSKI, T. H.; REYBROECK, W. Antimicrobials. In: International Dairy Federaltion standard (IDF/FIL). Monograph on residues and contaminants in milk and milk products. Brussels: IDF, 1997. Special Issue, p. 26-34
BYAKIKA, S.; MUKISA, I. M.; BYARUHANGA, Y. B.; MUYANJA, C. A Review of Criteria and Methods for Evaluating the Probiotic Potential of Microorganisms. Food Reviews International, n. 35, v. 5, p. 427-466. 2019.

CAO, H.; LIU, D.; MO, X.; XIE, C.; YAO, D. A fungal enzyme with the ability of aflatoxin B1 conversion: Purification and ESI-MS/MS identification. Microbiological Research, n. 166, p. 475-483. 2011.

CARNEIRO, M.; FERRAZ, T.; BUENO, M.; KOCH, B. E.; FORESTI, C.; LENA, V. P.; MACHADO, J. A.; RAUBER, J. M.; KRUMMENAAUER, E. C.; LAZAROTO, D. M. O uso of antimicrobials in a teaching hospital: a brief evaluation. Journal of the Brazilian Medical Association, v. 57, n. 4, p. 421-424. 2011.

CECCHINI, F.; MORASSUT, M.; SAIZ, J. C.; MORUNO, E. G. Anthocyanins enhance yeast's adsorption of Ochratoxin A during the alcoholic fermentation. European Food Research and Technology, v. 245, p. 309-314. 2019.

CHANG, X.; LIU, H.; SUN, J.; WANG, J.; ZHAO, C.; ZHANG, W.; ZHANG, J.; SUN, C. Zearalenone Removal from Corn Oil by an Enzymatic Strategy. Toxins, v. 12, p.117. 2020.

CHIESA, L. M.; DECASTELLI, L.; NOBILE, M.; MARTUCCI, F.; MOSCONI, G.; FONTANA, M.; CASTRICA, M.; ARIOLI, F.; PANSERI, S. Analysis of antibiotic residues in raw bovine milk and their impact toward food safety and on milk starter cultures in cheese-making process. LWT Food Science and Technology, n. 131, p. 109783. 2020.

CHIOCCHETTI, G. M.; JADAN-PIEDRA, C.; MONEDERO, V.; ZUNIGA, M.; VELEZ, D.; DEVESA, V. Use of lactic acid bacteria and yeasts to reduce exposure to chemical food contaminants and toxicity. Critical Reviews in Food Science and Nutrition, v. 59, p. 1534 - 1545. 2019.

CHO, S. W.; YIM, J.; SEO, S. W. Engineering Tools for the Development of Recombinant Lactic Acid Bacteria. Biotechnology Journal, n. 15, p. 1900344. 2020.

CHUANG, W. Y.; HSIEH, Y. C.; LEE, T-T. The Effects of Fungal Feed Additives in Animals: A Review. Animals, v. 10, n. 5, p. 805. 2020.

CIZEIKIENE, D.; JUODEIKIENE, G.; PASKEVICIUS, A.; BARTKIENE, E. Antimicrobial activity of lactic acid bacteria against pathogenic and spoilage microorganism isolated from food and their control in wheat bread. Food Control, v. 31, n. 2, p. 539-545. 2013.

CLSI - Clinical and Laboratory Standards Institute. M07-A9: Methods for Dilution Antimicrobial Susceptibility Tests for Bacteria That Grow Aerobically; Approved Standard-Ninth Edition, v. 32, n. 2. Pennsylvania: CLSI Wayne, 2012. 88p.

COFFEY, R.; CUMMINS, E.; WARD, S. Exposure assessment of mycotoxins in dairy milk. Food Control, v. 20, p. 239-249. 2009.

ČOLOVIĆ R, PUVAČA N, CHELI F, AVANTAGGIATO G, GRECO D, ĐURAGIĆ O, KOSJ, PINOTTI L. Decontamination of Mycotoxin-Contaminated Feedstuffs and Compound Feed.Toxins, v. 11, n. 11, p. 617. 2019.

CORREA, L.; FUKUSHIMA, A. R. Potential Antiviral Activity of Azithromycin: A Systematic Review. SanarMed, v. 3, p. 97-99. 2020.

CORTEZ, M. A. S.; DIAS, V. G.; MAIA, R. G.; COSTA, C. C. A. Physical and behavioural characteristics chemical and sensory analysis of pasteurised milk with added water, cheese whey, physiological whey and glucose serum. Revista do Instituto de Laticínios Cândido Tostes, v. 376, n. 65, p. 18-25. 2010.

da SILVA, F. C., CHALFOUN, S. M., BATISTA, L. R., SANTOS, C. AND LIMA, N. Polyphasic taxonomy for the identification of Aspergillus section flavi: a review. Electronic Journal Classroom in Focus, v.1, n.1, p. 18-40. 2015.

DALLMANN, E. P.; PASSOS, J. H.; TREIS, L. S.; MEDINA, M. C. R.; BASRG, M.

ZAMIGNAN, S. W.; CONSTANTE, T.; BATISTA, K. Z. S. Mycotoxins and their alarming reach in cattle farming: Review. PUBVET, v. 15, n. 9, p. 1-10. 2021.

DANIELI, B. B.; SCHOGO, A. L. Use of additives in ruminant nutrition: a review. Veterinary and Animal Science, v. 27, p. 1-13. 2020.

DAOU, R.; JOUBRANE, K.; MAROUN, R. G.; KHABBAZ, L. R.; ISMAIL, A.; EL KHOURY, A. Mycotoxins: Factors influencing production and control strategies. AIMS Agriculture and Food, n. 6, v. 1, p. 416 - 447. 2021.

de MIL, T.; DEVREESE, M.; DE BAERE, S.; VAN RANST, E.; EECKHOUT, M.; DE BACKER, P.; CROUBELS, S. Characterisation of 27 Mycotoxin Binders and the Relation with in Vitro Zearalenone Adsorption at a Single Concentration. Toxins, v. 7, p, 21-23. 2015.

de PAULA, C. E. R.; ALMEIDA, V G. K.; BORGES, R. M.; CASSELLA, R. J. Determination Spectrophotometric Analysis of Azithromycin in Pharmaceutical Formulations Using the Alizarin Reaction. Revista Virtual de Química, v. 11, n. 4, p.1081 -1096. 2019.

DELAVENNE, E.; MOUNIER, J.; DÉNIEL, F.; BARBIER, G.; LE BLAY, G. Biodiversity of antifungal lactic acid bacteria isolated from raw milk samples from cow, ewe and goat over one-year period. International Journal of Food Microbiology, v. 155, n. 3, p. 185 - 190. 2012

DENOBILE, M.; NASCIMENTO, E. S. Validation of a method for the determination of residues of the antibiotics oxytetracycline, tetracycline, chlortetracycline and doxycycline in milk by high-performance liquid chromatography. Brazilian Journal of Pharmaceutical Sciences, v. 40,n. 2, p. 209-2018. 2004.

DEVIKA, J. D.; APARNA, S.; JOHNSON, J. B.; SABU, T. Critical insights into antibiotic resistance transferability in probiotic Lactobacillus. Nutrition, v. 69, p. 110567. 2020.

DEVREESE M.; PASMANS F.; de BACKER P.; CROUBELS S. An in vitro model using the IPEC-J2 cell line for efficacy and drug interaction testing of mycotoxin detoxifying agents. Toxicology in Vitro, n. 27, p. 157- 163. 2013.

di CASTRO, I. C.; de OLIVEIRA, H. F.; MELLO, H. H . C.; MASCARENHAS, A. G. Mycotoxins in pig production. Portuguese Journal of Veterinary Sciences, v. 110, n 593-594, p. 6-13. 2015.

DOI, Y. Penicillins and β-lactamase inhibitors. In: MANDELL, G. L.; BENNET, J. E.; DOLIN,R. Mandell, Douglas, and Bennett's Principles and Practice of Infectious Diseases. 9. ed. Philadelphia: Elsevier Saunders, 2020. p. 251-267.

dos SANTOS, K. M. O.; VIEIRA, A. D. S.; ALONSO BURITI, F. C.; NASCIMENTO, J. C. F.; MELO, M. E. S.; BRUNO, L. M.; BORGES, M. F.; ROCHA, C. R. C.; LOPES, A. C. S.; FRANCO, B. D. G. M.; TODOROV, S. D. Artisanal coalho cheeses as source of beneficial Lactobacillus plantarum and Lactobacillus rhamnosus strains. Dairy Science and Technology, v. 95, n. 2, p. 209 - 230. 2015.

DOUILLARD, F . P . ; DE VOS, W . M . Biotechnology of health-promoting bacteria. Biotechnology Advances, v. 37, n. 6, p. 107369. 2019.

DUGGAR B. M. Aureomycin: a product of the continuing search for new antibiotics. Annals of the New York Academy of Sciences, v. 51, p. 177-181. 1948.

DZANTIEV, B.B.;BYZOVA,N.A.;URUSOV,A.E.;ZHERDEV,A.V.; Immunochromatographic methods in food analysis. Trends in Analytical Chemistry, v. 55, p. 81-93. 2014.

ELLIOT, C.T.; CONNOLLY, L., KOLAWOLE, O. Potential adverse effects on animal health

and performance caused by the addition of mineral adsorbents to feeds to reduce mycotoxin exposure. Mycotoxin Research, n. 36, p.115 - 126. 2019.
EL-NEZAMI, H.; KANKAANPÄÄ, P.; SALMINEN, A. J. Physicochemical alterations enhance the ability of dairy strains of lactic acid bacteria to remove aflatoxin from contaminated media. Food Protection, v, 61, p. 466-468. 1998.
EL-NEZAMI, H.; KANKAANPAA, P; SALMINEN, S; AHOKAS, J. Ability of Dairy Strains of Lactic Acid Bacteria to Bind a Common Food Carcinogen, Afatoxin B1. Food and Chemical Toxicology, n. 36, p. 321 - 326. 1998.
EL-NEZAMI, H.; POLYCHRONAKI, N.; SALMINEN, S.; MYKKÄNEN, H. Binding Rather Than Metabolism May Explain the Interaction of Two Food-Grade Lactobacillus Strains with Zearalenone and Its Derivative 'a -Zearalenol. Applied. Environmental Microbiology, n. 68, p.3545 -3 549. 2002.
EYANG, S. C.; ELIN, C. H.; SUNG, C. T.; EFANG, J. Y. Antibacterial Activities of Bacteriocins: Application in Foods and Pharmaceuticals. Frontier Microbiology, v.5, p. 241. 2014.FALKAUSKAS, R.; BAKUTIS, B.; JOVAIŠIENE, J.; VAICIULIEN, G.; GERULIS, G.; KERZIENE, S.; JACEVICIEN, I.; JACEVICIUS, E.; BALIUKONIENE, V. Zearalenone andIts Metabolites in Blood Serum, Urine, and Milk of Dairy Cows. Animals, v. 12, p. 1651. 2022.
FAO. Food and Agriculture Organisation of the United Nations. Biotechnology for agricultural development: proceedings of the FAO International Technical Conference on "agricultural biotechnologies in developing countries: options and opportunities in crops, forestry, livestock, fisheries and agro-industry to face the challenges of food insecurity and climate change" (ABCD-10). Rome: FAO, 2011. 569 p.
FAO. FOOD AND AGRICULTURE ORGANISATION OF UNITED NATIONS. Dairy and dairy products. In OECD-FAO agricultural outlook 2019-2028. Rome. 2019.
FAO. FOOD AND AGRICULTURE ORGANISATION OF UNITED NATIONS. Dairy market review. Rome. 2021.
FAO/WHO Food and Agriculture Organisation of the United Nations/World Health Organisation - CODEX ALIMENTARIUS. General standard for contaminants and toxins in food and feed CXS 193-1995. 2019. 65p.
FAO/WHO Working Group. Report on Drafting Guidelines for the Evaluation of Probiotics in Food: Guidelines for the Eval-uation of Probiotics in Food. London. 2002, p. 11.

FELTRIN, C. W.; MELLO, A. M. S.; SANTOS, J. G. R.; MARQUES, M. V.; SEIBEL, N. M.;FONTOURA, L. A. M. Quantification of sulfadimethoxine in milk by high-performance liquid chromatography. Química Nova, v. 30, n. 1, p. 80-82. 2007.
FERRO, E. S. Translational biotechnology: haemopressin and other intracellular peptides. Estudos Avançados, v. 24, n. 70, p. 109-121, 2010.
FLEMING, A. History and development of penicillinin. In: FLEMING, Alexander. Penicillin. Its practical application. London: Butterworth & Co, 1946. 2326 p.

FORSYTHE, S. J. Microbiology of Food Safety. Porto Alegre: Artmed, 2002. 422p.
FRANCISCATO, C.; LOPES, S. T. A.; SANTURIO, J. M.; WOLKMER, P.; MACIEL, R. M.;PAULA, M. T.; GARMATZ, B. C.; COSTA, M. M. Serum concentrations of minerals and liver and kidney functions of chickens intoxicated with aflatoxin and treated with sodium montmorillonite. Pesquisa Agropecuária Brasileira, v. 41, n. 11, p. 1573-1577. 2006.
FRUHAUF, S.; NOVAK, B.; NAGL, V.; HACKL, M.; HARTINGER, D.; RAINER, V.;

LABUDOVÁ, S.; ADAM, G.; ALESCHKO, M.; MOLL, W.D.; THAMHESL, M.; GRENIER,B. Biotransformation of the Mycotoxin ZENralenone to its Metabolites Hydrolyzed ZENralenone (HZEN) and Decarboxylated Hydrolyzed ZENralenone (DHZEN) Diminishes its Estrogenicity In Vitro and In Vivo. Toxins, v. 11, n. 8, p. 481. 2019.

FRY, D. E. Antimicrobial Peptides. Surgeon Infections, n. 19, p. 804-811. 2018.

FUCHS, S.; SONTAG, G.; STIDL, R.; EHRLICH, V.; KUNDI, M.; KNASMÜLLER, S. Detoxification of patulin and ochratoxin A, two abundant mycotoxins, by lactic acid bacteria. Food and chemical toxicology, v. 46, n. 4, p. 1398 - 1407. 2008.

GAJDA, A .; NOWACKA-KOZAK, E . ; GBYLIK-SIKORSKA, M .; POSYNIAK, A . Tetracycline antibiotics transfer from contaminated milk to dairy products and the effect of the skimming step and pasteurisation process on residue concentrations. Food Additives & Contaminants: Part A, n. 35, p. 66 - 76. 2018.

GALVÃO, L. C. Lactose intolerance. In: SPSP. Paediatric Society of São Paulo. Recommendations: Updating Conduct in Paediatrics. São Paulo: São Paulo Paediatric Society. 2012. 8p.

GAO, X.; MA, Q.; ZHAO, L.; LEI, Y.; SHAN, Y.; JI, C. Isolation of Bacillus subtilis: screening for aflatoxins B1, M1, and G1 detoxification. European Food Research and Technology, n. 232, v. 6, p. 957. 2011.

GARTLAN, W. A.; RAHMAN, S.; RETI, K. Benzathine Penicillin. In: StatPearls [Internet]. Treasure Island (FL): StatPearls Publishing. 2022.

GASTALHO, S.; SILVA, G. J.; RAMOS, P. Use of antibiotics in aquaculture and bacterial resistance: Impact on public health. Acta Farmacêutica Portuguesa, v. 3, n. 1, p. 29-45, 2014.

GAUTRET, P.; LAGIER, J.; PAROLA, P.; HOANG, V.; MEDDEB, L.; MAILHE, M.; DOUDIER, B.; COURJON, J.; GIORDANENGO, V.; VIEIRA, V. E.; DUPONT, H. T.; HONORÉ, S.; COLSON, P.; CHABRIÈRE, E.; LAS SCOLA, B.; ROALIN, J. M.; BROUQUI,P.; RAOULT, D. Hydroxychloroquine and azithromycin as a treatment of COVID-19: results of an open-label non-randomised clinical trial. International Journal of Antimicrobial Agents,v. 56, n. 1, p. 105949. 2020.

GERMANO, P. M. L.; GERMANO, M. I. S. Food Hygiene and Health Surveillance. 6. ed. São Paulo: Manole, 2019. 896 p.

GIL-SERNA, J.; VÁZQUEZ, C.; GONZÁLEZ-JAÉN, M. T.; PATIÑO, B. Mycotoxins Toxicology. In; BATT, C. A.; TORTORELLO, M. L. (org.) Encyclopedia of Food Microbiology, Cambridge: Academic Press. 2014. 887-892 p.

GONÇALVES, B. L.; ROSIM, R. E.; de OLIVEIRA, C. A. F.; CORASSIN, C. H. The in vitro ability of different Saccharomyces cerevisiae-based products to bind aflatoxin B1. Food Control, v. 47, p. 298-300. 2015.

GONÇALVES, B.; SANTANA, L.; PELEGRINI, P. Mycotoxins: A review of the main diseases triggered in the human and animal organism. Faciplac Health Journal, v. 4, n. 1, p. 1-12. 2017.

GONZÁLEZ, M. L.; SULYOK, M.; BARALLA, V.; DALCERO, A. M.; KRSKA, R.; CHULZE, S.; CAVAGLIERI, L. R. Evaluation of zearalenone, α-zearalenol, β-zearalenol, zearalenone 4-sulfate and β-zearalenol 4-glucoside levels during the ensiling process. World Mycotoxin Journal, n. 7, p. 291-295. 2014.

GRAHAM, C. E.; CRUZ, M. R.; GARSIN, D. A.; LORENZ, M. C. Enterococcus Faecalis

bacteriocin EntV Inhibits Hyphal Morphogenesis, Biofilm Formation, and Virulence of Candida Albicans. Proceedings of the National Academy of Sciences, v. 114, n. 17, p. 4507-4512. 2017.
GRAHAM, H.; MCCRACKEN, K. J. Yeasts in Animal Feeds. In: GARNSWORTHY P. C.; WISEMAN, J. Recent Advances in Animal Nutrition. Nottingham: University Press. 2005. p 169-211.
GRIFFITH, D. E. Mycobacterium abscessus and Antibiotic Resistance: Same As It Ever Was. Clinical Infectious Disease, v. 69, p. 1687. 2019.
GRUMET, L.; TROMP, Y.; STIEGELBAUER, V. The Development of High-Quality Multispecies Probiotic Formulations: From Bench to Market. Nutrient, n. 12, p. 2453. 2020.
GUIMARAES, A.; SANTIAGO, A.; TEIXEIRA, J. A.; VENANCIO, A.; ABRUNHOSA, L. Anti-aflatoxigenic effect of organic acids produced by Lactobacillus plantarum. International Journal of Food Microbiology, n. 264, p. 31-38. 2018.
GUO, C.; YUE, T. L.; HATAB, S.; YUAN, Y. H. Ability of inactivated yeast powder to adsorb patulin from apple juice. Journal of Food Protection, v. 75, n. 3, p. 585-590. 2012.
HAMAD, S. H. Factors Affecting the Growth of Microorganisms in Food In BHAT, R.; ALIAS, A. K; PALIYATH, G. (eds.). Progress in Food Preservation, 1 ed. New Jersey: John Wiley & Sons, 2012. 23 p.
HAMZA, Z.; EL-HASHASH, M.; ALY, S.; HATHOUT, A.; SOTO, E.; SABRY, B.; OSTROFF, G. Preparation and characterisation of yeast cell wall beta-glucan encapsulated humic acid nanoparticles as an enhanced aflatoxin B1 binder, Carbohydrates Polymeres, n. 203, p. 185-192. 2019.
HASKARD, C. A.; EL-NEZAMI, H. S.; KANKAANPÄÄ, P. E.; SALMINEN, S.; AHOKAS,J. T. Surface Binding of Aflatoxin B1 by Lactic Acid Bacteria. Applied. Environmental Microbiology, n. 67, p. 3086-3091. 2001.
HASSAN, H. F.; SAIDY, L.; HADDAD, R.; HOSRI, C.; ASMAR, S.; JAMMOUL, A.; JAMMOUL, R.; HASSAN, H.; SERHAN, M. Investigation of the effects of some processing conditions on the fate of oxytetracycline and tylosin antibiotics in the making of commonly consumed cheeses from the East Mediterranean. Veterinary World, n. 14, p. 1644 - 1649. 2021.
HASSAN, M. M.; ZAREEF, M.; XU, Y.; LI, H.; CHEN, Q. SERS-based sensor for mycotoxins detection: Challenges and improvements. Food Chemistry, v.344, p. 128652. 2021.
HASSAN, N. S.; JALIL, A. A.; HITAM, C. N. C.; VO, D. V. N.; NABGAN, W. Bio fuels and renewable chemicals production by catalytic pyrolysis of cellulose: a review. Environmental Chemistry Letters, v. 18. 1625-1648. 2020.
HATAB, S.; YUE, T. L.; MOHAMAD, O. Removal of patulin from apple juice using inactivated lactic acid bacteria. Journal of Applied Microbiology, n. 112, p. 892 - 899. 2012.
HATHOUT, A. S.; ALY, S. E. Biological detoxification of mycotoxins: A review. Annals of Microbiology, v. 64, p. 905 - 919. 2014.
HOLANDA, D. M.; KIM, S. W. Mycotoxin Occurrence, Toxicity, and Detoxifying Agents in Pig Production with an Emphasis on Deoxynivalenol. Toxins, v.13, n. 2, p. 171. 2021.
HUA, S. S.; BECK, J. J.; SARREAL, S. B.; GEE, W. The major volatile compound 2-phenylethanol from the biocontrol yeast, Pichia anomala, inhibits growth and expression of aflatoxin biosynthetic genes of Aspergillus flavus. Mycotoxin Research, n. 30, p. 71-78. 2014.

HUWIG, A.; FREIMUND, S.; KÄPPELI, O.; DUTLER, H. Mycotoxin detoxication of animal feed by different adsorbents. Toxicology Letters, v. 122, n. 2, p. 179 - 188. 2001.
IAL. Adolfo Lutz Institute. Physico-chemical methods for analysing food. 4. ed. São Paulo: Instituto Adolfo Lutz, 2008. 1020 p.
IARC Some naturally occurring substances: Food items and constituents, heterocyclic aromatic amines and mycotoxins IARC Monographs on the Evaluation of Carcinogenic Risks to Humans 1993, 56, 1-599.

IARC. World Health Organisation & International Agency for Research on Cancer. Some traditional herbal medicines, some mycotoxins, naphthalene and styrene in IARC Monographs on the Evaluation of Carcinogenic Risks to Humans v. 82. IARCPress, Lyon. 2002.
ISMAIL, A.; AKHTAR, S.; LEVIN, R. E.; ISMAIL, T.; RIAZ, M.; AMIR, M. Aflatoxin M1: Prevalence and decontamination strategies in milk and milk products. Critical Reviews in Microbiology, n. 42, p. 418-427. 2015.
ISMAIL, A.; GONÇALVES, B. L.; de NEEFF, D. V.; PONZILACQUA, B.; COPPA, CAROLINA F. S. C.; HINTZSCHE, H.; SAJID, M.; CRUZ, A. G.; CORASSIN, C. H.; OLIVEIRA, C. A. F. Aflatoxin in foodstuffs: Occurrence and recent advances in decontamination. Food Research International, n. 113, p. 74-85. 2018.
JARD, G.; LIBOZ, T.; MATHIEU, F.; GUYONVARC'H, A.; LEBRIHI, A. Review of mycotoxin reduction in food and feed: from prevention in the field to detoxification by adsorption or transformation. Food Additives & Contaminants: Part A: Chemistry, Analysis, Control, Exposure & Risk Assessment, v. 28, n. 11, p. 1590 - 1609. 2011.
JAY, J. M.; LOESSNER, M. J.; GOLDEN, D. A. Modern food microbiology, 7 ed. New York: Spring Science. 2005. 782p.
JENSEN, H.; GRIMMER, S.; NATERSTAD, K.; AXELSSON, L. In vitro testing of commercial and potential probiotic lactic acid bacteria. International Journal of Food Microbiology, n. 153, p. 216 - 222. 2012.
JI, C.; FAN, Y.; ZHAO, L. H. Review on biological degradation of mycotoxins. Animal Nutrition, n. 2, p. 127-133. 2016.
JIANG, Y.; OGUNADE, I. M.; KIM, D. H.; LI, X.; PECH-CERVANTES, A. A.; ARRIOLA, K. G.; OLIVEIRA, A. S.; DRIVER, J. P.; FERRARETTO, L. F.; STAPLES, C. R.; VYAS, D.;ADESOGAN, A. T. Effect of adding clay with or without a Saccharomyces cerevisiae fermentation product on the health and performance of lactating dairy cows challenged with dietary aflatoxin B1. Journal of Dairy Science, v. 101, n. 4, p. 3008-3020. 2018.
JITPAKDEE, J.; KANTACHOTE, D.; KANZAKI, H.; NITODA, T. Potential of lactic acid bacteria to produce functional fermented whey beverage with putative health promoting attributes. LWT Food Science and Technology, n. 160, p. 113269. 2022.
JOUANY, J. P.; YIANNIKOURIS, A.; BERTIN, G. The chemical bonds between mycotoxins and cell wall components of Saccharomyces cerevisiae have been identified. Archivos de Zootecnia, v.8, p. 26-50. 2005.
KAFLÉ, B. P. Chemical Analysis and Material Characterisation by Spectrophotometry. Amsterdam: Elsevier, 2019. 336 p.
KATZUNG, B. G.; TREVOR, A. J. Basic and clinical pharmacology 13 ed. Porto Alegre: AMGH. 2017. 1216 p.
KELLER, K. M.; OLIVEIRA, A. A.; ALMEIDA, T. X.; KELLER, L. A. M.; QUEIROZ, B.

D.; NUNES, L. M. T.; CAVAGLIERI, L. R.; ROSA, C. A. R. Effect of cell wall from yeast on the productive performance of broilers intoxicated with Aflatoxin B1. Brazilian Journal of Veterinary Medicine, v. 34, n. 2, p. 101-105. 2012.

KELLER, L. A. M., ABRUNHOSA, L., KELLER, K. M., ROSA, C. A. R., CAVAGLIERI, L.R. AND VENÂNCIO, A. Zearalenone and its derivatives α-zearalenol and β-zearalenol decontamination by Saccharomyces cerevisiae strains isolated from bovine forage. Toxins, v. 7, n. 8, p. 3297-3308. 2015.

KELLER, L. A. M.; ARONOVICH, M., KELLER, K. M.; CASTAGNA, A. A.; CAVAGLIERI, L. R.; ROSA, C. A. R. Incidence of Mycotoxins (AFB1 and AFM1) in Feeds and Dairy Farms from Rio de Janeiro State, Brazil. Veterinary. Medicine, v. 1, p. 29-35. 2016.

KLIS, F. M.; BOORSMA, A.; de GROOT, P. W. Cell wall construction in Saccharomyces cerevisiae. Yeast, v. 23, p. 185-202. 2006.

KÓSZEGI, T.; POÓR, M. Ochratoxin A: molecular interactions, mechanisms of toxicity and prevention at the molecular level. Toxins, v. 8, n. 111, p. 1-25. 2016.

KOVALSKY, P.; KOS, G.; NÄHRER, K.; SCHWAB, C.; JENKINS, T.; SCHATZMAYR, G.;SULYOK, M.; KRSKA, R. Co-occurrence of regulated, masked and emerging mycotoxins and secondary metabolites in finished feed and maize-an extensive survey. Toxins (Basel), v. 8,n. 363, p. 1-29. 2016.

KOVALSKY-PARIS, M. P.; SCHWEIGER, W.; HAMETNER, C.; STÜCKLER, R.; MUEHLBAUER, G. J.; VARGA, E.; KRSKA, R.; BERTHILLER, F.; ADAM, G. Zearalenone-16-O-glucoside: A New Masked Mycotoxin. Journal of Agricultural and Food Chemistry, v. 62, p. 1181-1189. 2014.

LANÇAS, P. M. Modern liquid chromatography and mass spectrometry: finally "compatible"?, Scientia Chromatographica, São Paulo, v. 1, n. 2, p. 35-61, 2009.

LANGDON A, CROOK N, DANTAS G. The effects of antibiotics on the microbiome throughout development and alternative approaches for therapeutic modulation. Genome Medicine, v. 8, n. 1, p. 1 - 16. 2016.

LEVINE, A. S.; FELLERS, C. R. Action of acetic acid on food spoilage microorganisms. Journal of Bacteriology, n. 39, v. 5, p. 499-515. 1940.

Li X, Li P, Zhang Q, Li R, Zhang W, Zhang Z, Ding X, Tang X: Multicomponent immunochromatographic assay for simultaneous detection of aflatoxin B1, ochratoxin A and zearalenone in agro-food. Biosens Bioelectron 2013, 49:426-432.

LI, J.; HUANG, J.; JIN, Y.; WU, C.; SHEN, D.; ZHANG, S.; ZHOU, R. Aflatoxin B1 degradation by salt tolerant Tetragenococcus halophilus CGMCC 3792. Food and Chemical Toxicology, n. 121, p. 430 - 434. 2018.

LI, P.; Sun. R.; YIN, R.; LAI, D.; WANG, M.; LIU, Y.; ZHOU, L. Detoxification of Mycotoxins through Biotransformation. Toxins, n. 12, v. 2, p. 121. 2020.

LI, T.; LI, L.; DU, F.; SUN, L.; SHI, J.; LONG, M.; CHEN, Z. Activity and Mechanism of Action of Antifungal Peptides from Microorganisms: A Review. Molecules, n. 26, p. 3438. 2021.

LI, T.; TENG, D.; MAO, R.; HAO, Y.; WANG, X.; WANG, J. A critical review of antibiotic resistance in probiotic bacteria. Food Research International, n. 136, p. 109571. 2020.

LIAO, Q.; RONG, H.; ZHAO, M.; LUO, H. CHU, Z.; WANG, R. Interaction between tetracycline and microorganisms during wastewater treatment: A review. Science of the Total Environment, n. 757, p. 143981. 2021.

LIMA, K. G. C.; KRUGER, M. F.; BEHRENS, J.; DESTRO, M. T.; LANDGRAF, M.; FRANCO, B. D. G. M. Evaluation of culture media for enumeration of Lactobacillus acidophilus, Lactobacillus casei and Bifidobacterium animalis in the presence of Lactobacillus delbrueckii subsp bulgaricus and Streptococcus thermophilus. LWT - Food Science and Technology, n. 42, p. 491-195. 2009.
LIU L, XIE M, WEI D. Biological Detoxification of Mycotoxins: Current Status and Future Advances. International Journal of Molecular Sciences, n. 23, v. 3, p. 10644. 2022.
LIU, J.; APPLEGATE, T. Zearalenone (ZEN) in Livestock and Poultry: Dose, Toxicokinetics, Toxicity and Estrogenicity. Toxins, v. 12, p. 377. 2020.
LIU, M.; ZHAO, L.; GONG, G.; ZHANG, L.; SHI, L.; DAI, J.; HAN, Y.; WU, Y.; KHALIL, M. M.; SUN, L. Invited review: Remediation strategies for mycotoxin control in feed. Journal of Animal Science and Biotechnology, n. 13, p. 19. 2022
LOPEZ, M. I.; PETTIS, J. S.; SMITH, I. B.; CHU, P. Multiclass determination and confirmation of antibiotic residues in honey using LC-MS/MS. Journal of Agricultural and Food Chemistry, v. 56, p. 1553-1559, 2008.
LU, Q.; LIANG, X.; & CHEN, F. Detoxification of Zearalenone by viable and inactivated cells of Planococcus sp. Food Control, v. 22, p. 191-195. 2011.
LU, Z. Y.; MA, Y. L.; ZHANG, J. T.; FAN, N. S.; HUANG, B. C.; JIN, R. C. A critical review of antibiotic removal strategies: Performance and mechanisms. Journal of Water Process Engineering, n. 38, p. 101681. 2020.
LUND, B. M.; EKLUND, T. Control of pH and use of organic acids. In LUND, B. M.; BAIRD- PARKER, T. C.; GOULD, G. W. (eds). The Microbiological Safety and Quality of Food, vol.1. Gaithersburg: Aspen Publishers. 2000. 1981p.
LUO, Y.; LIU, X. J.; LIU, Y.; HAN, Y. Q.; LI, J. K. Exogenous calcium ions enhance patulin adsorption capability of Saccharomyces cerevisiae. Journal of Food Protection, v. 82, n. 8, p. 1390-1397. 2019.
LUO, Y.; LIU, X.; YUAN, L.; LI, J. Complicated interactions between bio-adsorbents and mycotoxins during mycotoxin adsorption: Current research and future prospects Trends in Food Science & Technology, n. 96, p. 127-134. 2020.

LUO, Y.; WANG, J. G.; LIU, B.; WANG, Z. L.; YUAN, Y. H.; YUE, T. L. Effect of yeast cell morphology, cell wall physical structure and chemical composition on patulin adsorption. PLoS One, p. e0136045. 2015.
LUO, Y.; WANG, Z. L.; YUAN, Y. H.; ZHOU, Z. K.; YUE, T. L. Patulin adsorption of a superior microorganism strain with low flavour-affection of kiwi fruit juice. World Mycotoxin Journal, v. 9, n. 2, p. 195-203. 2016.
MAIA, P. P.; RATH, S. REYES, F. G. R. Antimicrobials in Foods of Plant Origin - A Review. Food and Nutrition Security, v. 16, n. 1, p. 49-64. 2009.
MAJDINASABA, M.; SHEIKH-ZEINODDIN, M.; SOLEIMANIAN-ZAD, S.; LI, P.; ZHANG, Q.; LI, X.; TANG, X. Ultrasensitive and quantitative gold nanoparticle-based immunochromatographic assay for detection of ochratoxin A in agro-products. Journal of Chromatography B, v. 974, p. 147-154. 2015.
MAKI, C. R.; THOMAS, A. D.; ELMORE, S. E.; ROMOSER, A. A.; HARVEY, R. B.; RAMIREZ, H. A.; PHILLIPS, T. D. Effects of calcium montmorillonite clay and aflatoxin exposure on dry matter intake, milk production, and milk composition. Journal of Dairy Science, v. 99, n. 2, p.1039-1046. 2016

MALIK, M.; BORA, J.; SHARMA, V. Growth studies of potentially probiotic lactic acid bacteria (Lactobacillus plantarum, Lactobacillus acidophilus, and Lactobacillus casei) in carrot and beetroot juice substrates. Journal of Food Processing and Preservation, v. 00, p. e14214. 2019.
MALLMANN, C. A.; DILKIN, P. Mycotoxins and Mycotoxicosis in Pigs. Santa Maria: Sociedade Vicente Pallotti. 2007. 238p.
MALLY, A.; SOLFRIZZO, M.; DEGEN, G.H. Biomonitoring of the mycotoxin Zearalenone: Current state-of-the art and application to human exposure assessment. Archives of Toxicology,v. 90, p. 1281-1292. 2016.
MARCONE, S.; BELTON, O.; FITZGERALD, D. J. Milk-derived bioactive peptides and their health promoting effects: a potential role in atherosclerosis. Brazilian Journal of Clinical Pharmacology, v. 83, n. 1, p. 152-162. 2017.
MARTIN, J. F.; ALVAREZ-ALVAREZ, R.; LIRAS, P. Penicillin-binding proteins, β-lactamases, and β-lactamase inhibitors in β-lactam-producing actinobacteria: self-resistance mechanisms. International Journal of Molecular Sciences, v. 23, n. 10, p. 5662. 2022.
MARTÍNEZ, J.; HERNÁNDEZ-RODRÍGUEZ, M.; MÉNDEZ-ALBORES, A.; TÉLLEZ-ISAÍAS, G.; JIMÉNEZ, E. M.; NICOLÁS-VÁZQUEZ, M. I.; RUVALCABA, R. M. Computational Studies of Aflatoxin B1 (AFB1): A Review. Toxins, n. 15, v. 2, p. 153. 2023.
MARTINEZ, M. P.; MAGNOLI, A. P.; GONZÁLEZ PEREYRA, M. L.; CAVAGLIERI, L. R. Probiotic bacteria and yeasts adsorb aflatoxin M1 in milk and degrade it to less toxic AFM1- metabolites. Toxicon, v. 172, p. 1 - 7. 2019.
MARTINEZ, R. C. R.; BEDANI, R.; SAAD, S. M. I. Scientific evidence for health effects attributed to the consumption of probiotics and prebiotics: an update for current perspectives and future challenges. British Journal of Nutrition, v. 144, n. 12, p. 1993 - 2015. 2015.
MASIMANGO, N.; REMACLE, J.; RAMAUT, J. L. The role of adsorption in the elimination of aflatoxin B1 from contaminated media. European Journal of Applied Microbiology Biotechnology, v. 6, p. 101-105. 1978.
MASSOUD, R.; DARANI, K. K.; SHARIFAN, A.; ASADI, G. H.; HADIANI, M. R. Mercury Biodecontamination from Milk by using L. acidophilus ATCC 4356. Journal of Pure and Applied Microbiology, v. 14, n. 4, p. 2313-1321. 2020.
MASTANJEVIĆ, K.; LUKINAC, J.; JUKIĆ, M.; ŠARKANJ, B.; KRSTANOVIĆ, V.;MASTANJEVIĆ, K. Multi-(myco)toxins in malting and brewing by-products. Toxins, v. 11, n. 30, p. 1-15. 2019.
MCAULIFFE, O. Symposium review: Lactococcus lactis from nondairy sources: their genetic and metabolic diversity and potential applications in cheese. Journal of Dairy Science, v. 101, p. 3597-3610. 2018.
MCCORMICK, S. P. Microbial detoxification of mycotoxins. Journal of Chemical Ecology, v. 39, p. 907-918. 2013.
MEUCCI, V.; RAZZUOLI, E.; SOLDANI, G.; MASSART, P. Mycotoxin detection in infant formula milks in Italy. Food Additives & Contaminants: Part A, v. 27, n. 1, p. 64-71. 2010.
MITCHELL, J. M.; GRIFFITHS, M. W.; McEWEN, S. A.; McNAB, W. B. YEE, A. J. Antimicrobial Drug Residues in Milk and Meat: Causes, Concerns, Prevalence, Regulations, Tests, and Test Performance. Journal of Food Protection, v. 61, n.6, p. 742-756. 1998.
MOEBUS, V. F.; PINTO, L. D. A.; KÖPTCKE, F. B. N.; KELLER, K. M.; ARONOVICH, M.; KELLER, L. A. M. In Vitro Mycotoxin Decontamination by Saccharomyces cerevisiae

Strains Isolated from Bovine Forage. Fermentation, v. 9, p. 585. 2023.
MORENO DE LEBLANC, A.; LEBLANC, J. G. Effect of probiotic administration on the intestinal microbiota, current knowledge and potential applications. World Journal of Gastroenterology, v. 20, n. 44, p. 16518 - 16528. 2014.
MOSTROM, M. S. Zearalenone. In GUPTA, R. (org.) Veterinary Toxicology; Basic and Clinical Principles, Cambridge: Academic Press, 2012. 1266-1271 p.
MUAZ, K.; RIAZ, M. Decontamination of aflatoxin m1 in milk through integration of microbial cells with sorbitan monostearate, activated carbon and bentonite. Journal of Animal and Plant Sciences, v. 31, n. 1, p. 235-245. 2021.
MUAZ, K.; RIAZ, M.; ROSIM, R. E.; AKHTAR, S.; CORASSIN, C. H.; GONÇALVES, B. L; OLIVEIRA, C. A. F. In vitro ability of nonviable cells of lactic acid bacteria strains in combination with sorbitan monostearate to bind to aflatoxin M1 in skimmed milk. LWT Food Science and Technology, v. 147, p. 111666. 2021.

NASCIMENTO, G. G. P.; MAESTRO, V.; CAMPOS, M. S. P. Occurrence of antibiotic residues in milk commercialised in Piracicaba, SP. Revista de Nutrição, v.14, n.2, p. 119-124. 2001.
NCBI. National Center for Biotechnology Information. 2022a. PubChem Compound Summary for CID 15558498, Aflatoxin M1. Retrieved March 24, 2023 from: https://pubchem.ncbi.nlm.nih.gov/compound/Aflatoxin-M1.
NCBI. National Centre for Biotechnology Information 2022b. PubChem Compound Summary for CID 5281576, ZENralenone. Retrieved March 24, 2023 from: https://pubchem.ncbi.nlm.nih.gov/compound/ZENralenone.
NDIAYE, S.; ZHANG, M.; FALL, M. AYESSOU, N. M.; ZHANG, Q.; LI, P. Current Review
of Mycotoxin Biodegradation and Bioadsorption: Microorganisms, Mechanisms, and Main Important Applications. Toxins, v. 14, n. 11, p. 729. 2022.
NI, Y.; HUANG, C.; KOKOT, S. A kinetic spectrophotometric method for the determination of ternary mixtures of reducing sugars with the aid of artificial neural networks and multivariate calibration. Analytica Chimica Acta, n.480, p. 53-65. 2003.
NIDERKORN V.; MORGAVI D. P.; ABOAB B.; LEMAIRE M.; BOUDRA H. Cell wall component and mycotoxin moieties involved in the binding of fumonisin B1, and B2, by lactic acid bacteria. Journal of Applied Microbiology, v. 106, p. 977 - 985. 2009.
NIH. NTM-CBI. Pub Chem: Compound Sumary. Azytromicin. 2022. Available at: https:// pubchem.ncbi.nlm.nih.gov/compound/447043. Accessed on: 28 Dec 2022.
NISHIMURA, J. Exopolysaccharides produced from Lactobacillus delbrueckii subsp. bulgaricus. Advances in Microbiology, v. 4, p. 1017-1023. 2014.
NISSAR, J.; AHAD, T.; NAZIR, F.; SALIM, R. Applications of biotechnology in food technology. International Journal of Engineering Technology Science, v. 4, n. 2, p. 300-306, 2017.
OLIVEIRA, A. A.; KELLER, K. M.; DEVEZA, M. V.; KELLER, L. A. M.; DIAS, E. O.; MARTINI-SANTOS, B. J.; AND ROSA, C.A.R., 2015. Effect of three different anti-mycotoxin additives on broiler chickens exposed to aflatoxin B1. Archivos de Medicina Veterinaria 47(2): 175-183.
OLIVEIRA, C. A. P.; GERMANO, P. M. L. Evaluation of the performance of the enzyme-linked immunosorbent assay (ELISA) method in reconstituted milk powder experimentally

contaminated with aflatoxin M1. Revista Saúde Pública, v. 30, n. 6, p.542-548. 1996.
OLIVEIRA, M. N. Tecnologia de produtos lácteos funcionais. São Paulo: Atheneu Publishing House. 2009. 384p.
OLIVER, M. E.C.; FAUSTO, M. C; SARAICA, L. H. G.; de CARVALHO, L. M.; PINTO, R.mycotoxins and mycotoxicosis in pig farming: literature review. Nutritime, v. 17, n. 2. 2020
UN. United Nations Organisation. Convention on Biological Diversity. Rio de Janeiro, 1992. Disponívelem: https://treaties.un.org/doc/Treaties/1992/06/19920605%2008-44%20PM/Ch_XXVII_08p.pdf. Accessed on: 02 Feb. 2022.
ORDÓÑEZ, J. A. Food Technology. Porto Alegre: Artmed. 2005. 280 p.
OSWALD, S.; KARSUNKE, X. Y.; DIETRICH, R.; MÄRTLBAUER, E.; NIESSNER, R.; KNOPP, D. Automated regenerable microarray-based immunoassay for rapid parallel quantification of mycotoxins in cereals. Analytical and Bioanalytical Chemistry, v. 405, p. 6405-6415. 2013.
PANCIERE, B. M.; RIBEIRO, L. F. Detection and occurrence of fraud in fluid milk or derivatives. Revista GETEC, Gestão, tecnologia e ciências, v. 10, n. 26, p. 1-17. 2021.
PANG, X.; SONG, X.; CHEN, M.; TIAN, S.; LU, Z.; SUN, J.; LI, X. LU, Y.; YUK, H. G. Combating biofilms of foodborne pathogens with bacteriocins by lactic acid bacteria in the food industry. Comprehensive Reviews in Food Science and Food Safety, v. 21, n. 2, p. 1657 - 1676. 2022
PARNHAM, M. J.; HABER, V. E.; GIAMARELLOS-BOURBOULIS, E. J.; PERLETTI, G.; VERLEDEN, G. M.; VOS, R. Azithromycin: Mechanisms of action and their relevance for clinical applications, Pharmacology & Therapeutics, v. 143, n. 2, p. 225-245. 2014.
PASSOS, M. L. C.; SARAIVA, M. L. M. F. S. Detection in UV-visible spectrophotometry: Detectors, detection systems, and detection strategies. Measurement, v. 135, p. 896-904. 2019.
PELES, F.; SIPOS, P.; KOVÁCS, S.; GYORI, Z.; PÓCSI, I.; PUSZTAHELYI, T., 2021. Biological control and mitigation of aflatoxin contamination in commodities. Toxins 13: 104.
PELTONEN, K. D.; EL-NEZAMI, H.; HASKARD, C.; AHOKAS, J.; SALMINEN, S. Aflatoxin B1 Binding by Dairy Strains of Lactic Acid Bacteria and Bifidobacteria. Journal of Dairy Science, v. 84, p. 2152-2156. 2001.
PEREIRA-MAIA, E. C.; SILVA, P. P.; ALMEIDA, W. B.; SANTOS, H. F.; MARCIAL, B. L.; RUGGIERO, R. GUERRA, W. Tetracyclines and glycylcyclines: an overview. Química nova, v. 33, n. 3, p. 700-706. 2010.
PEREYRA, C. M.; GIL, S.; CRISTOFOLINI, A.; BONCI, M.; MAKITA, M.; MONGE, M. P.; MONTENEGRO, M. A.; CAVAGLIERI, L. R. The production of yeast cell wall using an agroindustrial waste influences the wall thickness and is implicated on the aflatoxin B1 adsorption process. Food Research International, n. 111, p. 306-313. 2018
PEREYRA, M. L. G.; DOGI, C.; LISA, A.; WITTOUCK, P.; ORTIZ, M.; ESCOBAR, F.; BAGNIS, G.; YACIUK, R.; POLONI, L.; TOREES, A.; DALCERO, A. M.; CAVAGLIERI, L. R. Genotoxicity and cytotoxicity evaluation of probiotic Saccharomyces cerevisiae RCRC016: A 60-day subchronic oral toxicity study in rats. Journal of Applied Microbiology, v. 117, p. 824-833. 2014.
PETERIA, Z.; TERENB, J.; VAGVOLGYIA, C.; VARGAA, J. Ochratoxin degradation and adsorption caused by astaxanthin-producing yeasts. Food Microbiology, v. 24, p. 205-210. 2007.

PETROVA, P .; ARSOV, A .; TSVETANOVA, F .; PARVANOVA-ANCHEVA, T .; VASILEVA, E.; TSIGORIYNA, L.; PETROV K. The Complex Role of Lactic Acid Bacteria in Food Detoxification. Nutrients, v. 14, n. 10, p. 2038. 2022.

PETRUZZI, L.; CORBO, M. R.; SINIGAGLIA, M.; BEVILACQUA, A. Ochratoxin A removal by yeasts after exposure to simulated human gastrointestinal conditions. Journal of Food Science, v. 81, n. 11, p. 2756-2760. 2016.

PFLIEGLER, W. P.; PUSZTAHELYI, T.; PÓCSI, I. Mycotoxins-prevention and decontamination by yeasts. Journal of Basic Microbiology, v. 55, p. 805-818. 2015.

PINTO, A.; BARBOSA, J.; ALBANO, H.; ISIDRO, J.; TEIXEIRA, P. Screening of Bacteriocinogenic Lactic Acid Bacteria and Their Characterisation as Potential Probiotics. Microorganisms, v. 8, p. 393. 2020.

PIZZOLITTO, R. P.; SALVANO, M. A.; DALCERO, A. M. Analysis of fumonisin B1 removal by microorganisms in co-occurrence with aflatoxin B1 and the nature of the binding process. International Journal of Food Microbiology, v. 156, p. 214-221. 2012.

POLONI, V.; DOGI, C.; PEREYRA, C. M.; FERNÁNDEZ JURI, M. G.; KOHLER, P.; ROSA,
C. A. R.; DALCERO, A. M.; CAVAGLIERI, L. R. Potentiation of the effect of a commercial animal feed additive mixed with different probiotic yeast strains on the adsorption of aflatoxin B1. Food Additives & Contaminants: Part A, v. 32, n. 6, p. 970-976. 2015.

PRADO C. K.; MACHINSKI JUNIOR M. Analytical methodology for the determination of tetracycline residues in milk: a review. Revista do Instituto Adolfo Lutz, v. 70, n. 4. P. 448-456. 2011.

PRAPAPANPONG, J., UDOMKUSONSRI, P., MAHAVORASIRIKUL, W., CHOOCHUAY,S.; TANSAKUL. S. In vitro studies on gastrointestinal monogastric and avian models to evaluate the binding efficacy of mycotoxin adsorbents by liquid chromatography-tandem mass spectrometry. Journal of Advanced Veterinary and Animal Research, v.6, n. 1, p. 125-132. 2019

QIU Z.; FANG C.; GAO Q.; BAO J. A short-chain dehydrogenase plays a key role in cellulosic D-lactic acid fermentability of Pediococcus acidilactici. Bioresource Technology, v. 297, p. 122473. 2019.

QUINTANILLA, P.; BELTRÁN, M. C.; MOLINA, A.; ESCRICHE, I.; MOLINA, M. P. Characteristics of ripened Tronchón cheese from raw goat milk containing legally admissible amounts of antibiotics. Journal of Dairy Science, v. 102, p. 2941-2953. 2019.

RAI, A.; DAS, M.; TRIPATHI, A. Occurrence and toxicity of a fusarium mycotoxin, zearalenone. Critical Reviews in Food Science and Nutrition, v. 60, p. 2710-2729. 2019.

RALLA K.; SOHLING U.; RIECHERS D.; KASPER C.; RUF F.; SCHEPER T. Adsorption and separation of proteins by a smectite clay mineral. Bioprocess Biosystem Engineering, v. 33,p. 847-861. 2010.

RAWOOF, S. A. A.; KUMAR, P. S.; VO, D. V. N.; DEVARAJ, K.; MANI, Y.; DEVARAJ, T.; SUBRAMANIAN, S. Production of optically pure lactic acid by microbial fermentation: a review. Environmental Chemistry Letters, v. 19, p. 539-556. 2021.

RAY B.; BHUNIA A. Fundamental food microbiology, 5 ed. Boca Raton: CRC Press. 2013. 663p.

REIS, T. L.; DILELIS, F.; FASSANI, E. J.; CALIXTO, L. F. L. Aluminosilicates in the poultry feed: literature review. Research, Society and Development, v. 9, n. 8. 2020.

REN, X.; ZHANG, Q.; ZHANG, W.; MAO, J.; LI, P. Control of Aflatoxigenic Molds by Antagonistic Microorganisms: Inhibitory Behaviors, Bioactive Compounds, Related Mechanisms, and Influencing Factors. Toxins, n. 12, p. 24. 2020.
RIITER, J. M.; FLOWER, R.; HENDERSON, G.; LOKE, Y. K.; MacEWAN, D.; RANG, H. P. Rang & Dale Farmacologia 9 ed. Rio de Janeiro: GEN Guanabara Koogan. 2020. 808 p.
RODRIGUES, M. X.; DALL'AGNOLB, L.; BITTENCOURTA, J. V. M. Survey of the occurrence of antibiotic residues in raw milk produced in the Campos Gerais region, Paraná. UNOPAR Científica Ciência Biologicas e Saúde, n. 14, v. 4, p. 237-240, 2012.
RODRIGUES, R. O.; RODRIGUES, R. O.; LEDOUX, D. R.; ROTTINGHAUS, G. E.; BORUTOVA, R.; AVERKIEVA, O.; MCFADDEN, T. B. Feed additives containing sequestrant clay minerals and inactivated yeast reduce aflatoxin excretion in milk of dairy cows. Journal of Dairy Science, v. 102, n. 7, p. 1-10.
ROLFE, C.; DARYAEI, H. Intrinsic and Extrinsic Factors Affecting Microbial Growth in Food Systems In DEMIRCI, A.; FENG, H.; KRISHNAMURTHY, K. (eds.) Food Safety Engineering, 1 ed. Cham, Springer,2020. 753 p.
RONQUILLO, M. G.; HERNANDEZ, J. C. A. Antibiotic and synthetic growth promoters in animal diets: Review of impact and analytical methods. Food Control, v. 72, part B, p. 255-267. 2017
ROSSI, P; NUNES, J. K; RUTZ, F; REIS, J. S; BOURSHEIDT, D; ROCHA, A. A; MORAES,P. V. D; ANCIUTI, M. A. Glucomannan and selenium in chickens fed diets containing aflatoxins. Archivos de Zootecnia, v. 62, n. 237, p.33-43.2013.
RUSSO, P.; ARENA, M. P.; FIOCCO, D.; CAPOZZI, V.; DRIDER, D.; SPANO, G. Lactobacillus plantarum with broad antifungal activity: A promising approach to increase safety and shelf-life of cereal-based products. International Journal of Food Microbiology, n. 247, p.48-54. 2017.
RUTHVEN, D. M. Principles of Adsorption and Adsorption processes. New York: John Wiley & Sons, 1984. 432 p.

SACHI, S .; FERDOUS, J.; SIKDER, M. H.; HUSSANI, S. M. A. K. Antibiotic residues in milk: Past, present, and future. Journal of Advanced Veterinary and Animal Research, v. 6, n. 3, p. 315-332. 2019.
SADIQ, F. A.; YAN, B.; TIAN, F.; ZHAO, J. Lactic Acid Bacteria as Antifungal and Anti-Mycotoxigenic Agents: A Comprehensive Review. Comprehensive Reviews in Food Science and Food Safety, n. 18, p. 1403 - 1436. 2019.
SAEKI, E. K.; FARHAT, L. P.; PONTES, E. A. Efficiency of glycerol and skimmed milk cryoprotectants for freezing microorganisms. Acta Veterinaria Brasilica, v. 9, n. 2, p. 195 - 198. 2015.
SANGMANEE, P.; HONGPATTARAKERE, T. Inhibitory of multiple antifungal components produced by Lactobacillus plantarum K35 on growth, aflatoxin production and ultrastructure alterations of Aspergillus flavus and Aspergillus parasiticus. Food Control, n. 40, p. 224-233. 2014.
SANKAR, R. P; MADHURI, B.; LAKSHMI, N. A.; POOJA, A.; SAI, B. M.; SURESH, K. BABU, S. P. Selected HPLC Applications - Quick Separation Guide: A Review. International Journal of Pharmaceutical Sciences Review and Research, v. 60, n. 2, p. 13-20. 2020.
SARLAK, Z.; ROUHI, M.; MOHAMMADI, R.; KHAKSAR, R.; MORTAZAVIAN, A. M.; SOHRABVANDI, S.; GARAVAND, F. Probiotic biological strategies to decontaminate

aflatoxin M 1 in a traditional Iranian fermented milk drink (Doogh). Food Control, n.71, p. 152 - 159. 2017.
SAVAIANO, D. A.; HUTKINS, R. W. Yogurt, cultured fermented milk, and health: A systematic review. Nutrition reviews 2021. 79(5), 599-614
SCHULZ, P.; RIZVI, S. S. H. Hydrolysis of Lactose in Milk: Current Status a Future Products.Food Reviews International. 2021.
SCOTT, P. M. Natural Toxins. In: CUNNIF, P. (ed). Official Methods of Analysis of Association of Official Analytical Chemists. Gaithersburg, Maryland, 1997, p. 970.
SHAO, S.; JIA, X.; ZHANG, J.; MENG, J.; WU, H.; DUAN, H.; TU, X. Multi-residual analysis of 16 beta-agonists in pig liver, kidney and muscle by ultra performance liquid chromatography tandem mass spectrometry. Food Chemistry, v. 114, n. 3, p. 1115-1121, 2009.
SHAO, X.; XU, B.; CHEN, C.; LI, P.; LUO, H. The function and mechanism of lactic acid bacteria in the reduction of toxic substances in food: a review. Critical Reviews in Food Science and Nutrition, v. 62, n. 21, p. 5950-5963. 2022.
SHARMA, V.; ASERI, G.; BHAGWAT, P.; JAIN, N.; RANVEER, R. C. Production, Purification and Characterisation of a Novel Bacteriocin Produced by Bacillus subtilis VS Isolated from Mango (Mangifera indica L.). Brazilian Archives of Biology and Technology, v. 64, p. e21190749. 2021.

SHEHATA, M. G.; EL SOHAIMY, S. A.; EL-SAHN, M. A.; YOUSSEF, M.M. Screening of isolated potential probiotic lactic acid bacteria for cholesterol lowering property and bile salt hydrolase activity. Annals of Agricultural Sciences, v. 61, p. 65-75. 2016.
SHETTY, P. H., & JESPERSEN, L. Saccharomyces cerevisiae and lactic acid bacteria as potential mycotoxin decontaminating agents. Trends in Food Science & Technology, v. 17, p. 48-55. 2006.
SHUNDO, L.; RUVIERI, V.; NAVAS, S. A.; SABINO, M. Optimisation of the determination of aflatoxin M1 in milk using an immunoaffinity column and thin layer chromatography. Revista do Instituto Adolfo Lutz, v. 63, n. 1, p. 43-48. 2004.
SHUNDO, L.; SABINO, M. Aflatoxin m1 in milk by immunoaffinity column cleanup with TLC/HPLC determination. Brazilian Journal of Microbiology, v: 37, p.164-167,2006.
SIDDIQUI, M. S.; THODEY, K.; TRENCHARD, I.; SMOLKE, C. D. Advancing secondary metabolite biosynthesis in yeast with synthetic biology tools. FEMS Yeast Research, n. 12, p. 144-170. 2012.
SIGNORINI, M. L.; GAGGIOTTI, M.; MOLINERI, A.; CHIERICATTI, C. A.; BASÍLICO M. L. Z.; BASÍLICO, J. C.; PISANI, M. Exposure assessment of mycotoxins in cow's milk in Argentina. Food and Chemical Toxicology, v. 50, p. 250-257. 2012.
SILVA, C. J.; TEJADA, T. S.; TIMM, C. D. Resistance of Salmonella isolated from humans and chickens to antimicrobials. Revista Brasileira de Higiene e Sanidade Animal, v. 8, n. 4, p. 120-131, 2014.
SILVA, H. O.; VIDAL, A. M. C.; NETTO, A. S. Technologies applied to milk and dairy products. In: NETTO, A. S.; VIDAL, A. M. C. (org.) Obtaining and processing milk and dairy products. São Paulo: Faculty of Animal Science and Food Engineering, University of São Paulo. 2018. 2020 p.
SIQUEIRA-BATISTA, R.; GOMES, A. P.; SANTANA, L. A.; GELLER, M. Clinical use of antimicrobials: an update. Brazilian Journal of Medicine, v. 68, p. 154-157. 2011.

SIROLI, L.; PATRIGNANI, F.; SERRAZANETTI, D.I.; PAROLIN, C.; PALOMINO, R.A.Ñ;VITALI, B.; LANCIOTTI, R. Determination of Antibacterial and Technological Properties of Vaginal Lactobacilli for Their Potential Application in Dairy Products. Frontiers in Microbiology, v. 8, p. 1. 2018.
SMITH, B. P. Disorders caused by toxic substances. In: GALEY, F. D. (org.) Large Animal Internal Medicine. 3rd ed. Barueri: Manole, 2006. 1627-1630 p.
SNIFFEN, J. C.; MCFARLAND, L. V.; EVANS, C. T.; GOLDSTEIN, E. J. C. Choosing an appropriate probiotic product for your patient: An evidence based practical guide. PLoS ONE, v. 13, n. 12, p. e0209205. 2018.
SOLTANI, S.; HAMMAMI, R.; COTTER, P. D.; REBUFFAT, S.; SAID, L. B.; GAUDEAU, H.; BÉDARD, F.; BIRON, E.; DRIDER, D.; FLISS, I. Bacteriocins as a new generation of antimicrobials: toxicity aspects and regulations. FEMS Microbiology Reviews, v. 45, p. 1 - 24. 2021.
SONG, S.; LIU, N.; ZHAO, Z.; EDIAGE, E. N.; WU, S.; SUN, C.; SAEGER, S.; WU, A. Multiplex lateral flow immunoassay for mycotoxin determination. Analytical Chemistry, v. 86,p. 4995-5001. 2014.
SOUTO, P. C. M. C.; AUGUSTO, L.; GREGÓRIO, M. C. D.; OLIVEIRA, C. A. F. Principal mycotoxicosis in pigs. Veterinária e Zootecnia, v. 24, n. 3, p. 480-494. 2017.
SPINELLI, F. R.; CECCARELLI, F.; DI FRANCO, M.; CONTI, F. To consider or not antimalarials as a prophylactic intervention in the SARS-CoV-2 (Covid-19) pandemic. Annals of the rheumatic diseases, v. 79, n. 5, p. 666-667. 2020.
SRIVASTAVA, R. K.; SHETTI, N. P.; REDDY, K. R.; AMINABHAVI, T. M. Biofuels, biodiesel and biohydrogen production using bioprocesses: a review. Environmental Chemistry Letters, v.18, 1049-1072. 2020.
SUN, Y.; HU, X.; ZHANG, Y.; YANG, J.; WANG, F.; WANG, Y.; DENG, R.; ZHANG, G. Development of an Immunochromatographic Strip Test for the rapid detection of zearalenone in corn. Journal of Agricultural and Food Chemistry, v. 62, p. 11116-11121. 2014.
SURI, S.; KUMAR, V.; PRASAD, R.; TANWAR, B.; GOYAL, A.; KAUR, S.; GAT, Y.; KUMAR, A.; KAUR, J.; SINGH. Considerations for development of lactose-free food. Journal of Nutrition & Intermediary Metabolism, v. 15, p. 27-34. 2019.
TAYEL, A. A.; EL-TRAS, W. F.; MOUSSA, S. H.; EL-AGAMY, M. A. Antifungal action of Pichia anomala against aflatoxigenic Aspergillus flavus and its application as a feed supplement. Journal of Science Food Agriculture, n. 93, p. 3259-3263. 2013.
TIAN, L.; KHALIL, S.; BAYEN, S. Effect of thermal treatments on the degradation of antibiotic residues in food. Critical Reviews in Food Science and Nutrition, v. 57, p. 3760 - 3770. 2017.
TIMOTHY, B.; ILIYASU, A. H.; ANVIKAR, A. R. Bacteriocins of Lactic Acid Bacteria and Their Industrial Application. Current Topics in Lactic Acid Bacteria and Probiotics, v. 7, n. 1, p. 1-13. 2021.TONON, K. M.; SAVI, G. D.; SCUSSEL, V. M. Application of a LC-MS/MS method for multimycotoxin analysis in infant formula and milkbased products for young children commercialised in Southern Brazil. Journal of Environmental Science and Health, Part B, v. 0,n. 0, p. 1-7. 2018.
TORTORA, G. J.; FUNKE, B. R.; CASE, C. L. Microbiologia, 12 ed. Porto Alegre: Artmed Editora. 2017. 962p.
TREIBER, F.M.; BERANEK-KNAUER, H. Antimicrobial Residues in Food from Animal

Origin-A Review of the Literature Focusing on Products Collected in Stores and Markets Worldwide. Antibiotics, v. 10, p. 534.

TROMBETE, P. M.; SANTOS, R. R.; SOUZA, A. L. R. Antibiotic residues in Brazilian milk: a review of studies published in recent years. Revista Chilena de Nutrición, v. 41, n. 2, p. 191- 195, 2014.

TRUCKSESS, M. W. Rapid analysis (thin layer chromatographic and immunochemical methods) for mycotoxins in foods and feeds. In: de KOE, W. J.; SAMSOM, R. A.; VAN EGMOND, H. P.; GILBERT, J.; SABINO, M. (eds). Ponsen&Looyen, Wageningen, The Netherlands, 2001, p.29-40.

TSIPLAKOU, E.; ANAGNOSTOPOULOS, C.; LIAPIS, K.; HAROUTOUNIAN, S. A.; ZERVAS, G. Determination of mycotoxins in feedstuffs and 'ruminants milk using an easy and simple LC-MS/MS multiresidue method. Talanta, v. 130, p. 8-19. 2014.

TURNA, N. S.; WU, F. Aflatoxin M1 in milk: A global occurrence, intake, & exposure assessment. Trends in Food Science and Technology, v. 110, p. 183-192. 2021.

TYUTKOV, N.; ZHERNYAKOVA, A.; BIRCHENKO, A.; EMINOVA, E.; NADTOCHII, L.BARANENKO, D. Probiotics viability in frozen food products. Food Bioscience, v. 50, p. 101996. 2022.

UEBERSCHÄR, K.-H.; BREZINA, U.; DÄNICKE, S. Zearalenone (ZEN) and ZEN metabolites in feed, urine and bile of sows: Analysis, determination of the metabolic profile and evaluation of the binding forms. Applied Agricultural and Forestry Research, v. 1, p. 21-28. 2016.

URUSOV, A. E.; ZHERDEV, A. V.; PETRAKOVA, A. V.; SADYKHOV, E. G.; KOROLEVA, O. V.; DZANTIEV, B. B. Rapid multiple immunoenzyme assay of mycotoxins.Toxins, v. 7, p. 38-254. 2015.

UYENO, Y.; SHIGEMORI, S.; SHIMOSATO, T. Effect of probiotics/prebiotics on cattle health and productivity. Microbes and environments, v. 30, n. 2, p. 126-132.

VAZ, A.; SILVA, A. C. C.; RODRIGUES, P.; VENÂNCIO, A. Detection Methods for Aflatoxin M1 in Dairy Products. Microorganisms, v. 8, n. 2, p. 246. 2020.

VICENTE, D.; PÉREZ-TRALLERO, E. Tetracyclines, sulphamides y metronidazole. Enfermedades Infecciosas y Microbiología Clinica, v. 28, n. 2, p. 122-130. 2010.

VILA-DONAT, P.; MARÍN, S.; SANCHIS, V.; RAMOS, A.J. A review of the mycotoxin adsorbing agents, with an emphasis on their multi-binding capacity, for animal feed decontamination.Food and Chemical Toxicology, v. 114, p. 246-259. 2018.

VIRTO, M.; SANTAMARINA-GARCÍA, G.; AMORES, G.; HERNÁNDEZ, I. Antibiotics in Dairy Production: Where Is the Problem? Dairy, v. 3, p. 541-564. 2022.

WALSTRA, P.; WOUTERS, J.T.M.; GEURTS, T.J. Dairy Science and Technology. 2nd ed. Boca Raton: CRC Press, 2006.

WALTER, P. P.; TUNDE, P.; ISTVAN, P. Mycotoxins-prevention and decontamination by yeast. Journal of Basic Microbiology, v. 55, p. 805-818. 2015.

WAN, M. L. Y.; FORSYTHE, S. J.; EL-NAZAMI, H. Probiotics interaction with foodborne pathogens: a potential alternative to antibiotics and future challenges. Reviews in Food Science and Nutrition, v. 59, n. 20, p. 3320 - 3333. 2019.

WANG G.; MIAO Y.; SUN Z.; ZHENG S. Simultaneous adsorption of aflatoxin B1 and zearalenone by mono- and di-alkyl cationic surfactants modified montmorillonites. Journal of Colloid Interface Science, n. 511, p. 67 - 76. 2018.

WANG, Y. K.; YAN, Y. X.; JI, W. H.; WANG, H. A.; LI, S. Q.; ZOU, Q.; SUN, J. H. Rapid simultaneous quantification of Zearalenone and Fumonisin B1 in corn and wheat by Lateral Flow Dual Immunoassay. Journal of Agricultural and Food Chemistry, v. 61, p. 5031-5036. 2013.

WANG, J.; XIE, Y. Review on microbial degradation of ZENralenone and aflatoxins. Grain & Oil Science and Technology, v.3, n. 3, p. 117-125. 2020.

WANG, L.; WANG, Z. L.; YUAN, Y. H.; CAI, R.; NIU, C.; YUE, T. L. Identification of key factors involved in the biosorption of patulin by inactivated lactic acid bacteria (LAB) cells. PLoS One, 1-19. 2015b.

WANG, L.; YUE, T. L.; YUAN, Y. H.; WANG, Z. L.; YE, M. Q.; CAI, R. A new insight into the adsorption mechanism of patulin by the heat-inactive lactic acid bacteria cells. Food Control, n. 50, p.104-110. 2015a.

WANG, Y.; WU, J.; LU, M.; SHAO, Z.; HUNGWE, M.; WANG, J.; BAI, X.; XIE, J.; WANG,Y.; GENG, W. Metabolism Characteristics of Lactic Acid Bacteria and the Expanding Applications in Food Industry. Frontiers in Bioengineering and Biotechnology, v. 9, p. 612285. 2021.

WHO. WORLD HEALTH ORGANISATION & FOOD AND AGRICULTURE, ORGANISATION OF THE UNITED NATIONS. Safety evaluation of certain mycotoxyns in food (Who Food Additivies Series 47/FAO Food and Nutrition Paper). Geneva, 2001.

WHO/FAO. WORLD HEALTH ORGANISATION & FOOD AND AGRICULTURE, ORGANISATION OF THE UNITED NATIONS. Guidelines for the evaluation of probiotics in food. FAO/WHO Working Group on Drafting Guidelines for the Evaluation of Probiotics in Food, p. 1 - 11. 2002.

WIDYASTUTI, Y.; FEBRISIANTOSA, A.; TIDONA, F. Health-Promoting Properties of Lactobacilli in Fermented Dairy Products. Frontiers in Microbiology, v. 12, p. 673890. 2021.

WIEËRS, G.; BELKHIR, L.; ENAUD, R.; LECLERCQ, S.; de FOY, J. M. P.; DEQUENNE, I.; de TIMARY, P.; CANI, P. D. How Probiotics Affect the Microbiota. Frontiers in Cellular and Infection Microbiology, v. 9, p. 454. 2020.

XAVIER-SANTOS, D.; BEDANI, R.; LIMA, E. D.; SAAD, S. M. I. Impact of probiotics and prebiotics targeting metabolic syndrome. Journal of Functional Foods, v. 64, p. 103666. 2020.

XIONG, J. L.; WANG, Y. M.; ZHOU, H. L.; LIU, J. X. Effects of dietary adsorbent on milk aflatoxin M1 content and the health of lactating dairy cows exposed to long-term aflatoxin B1 challenge. Journal of Dairy Science, v. 101, p. 8944-8953. 2018.

YANG, C.; SONG, G.; LIM, W. Effects of mycotoxin-contaminated feed on farm animals. Journal of Hazardous Materials, v. 389, p. 122087. 2020.

YIANNIKOURIS, A.; POUGHON, L.; CAMELEYRE, X.; DUSSAP, C. G.; FRANCOIS, J.; BERTING, G.; JOUANY. J. P. A novel technique to evaluate interactions between Saccharomyces cerevisiae cell wall and mycotoxins: Application to zearalenone. Biotechnology Letters, v. 25, n. 10, p. 783-789, 2003.

YIANNIKOURIS, A.; FRANCOIS, J.; POUGHON, L.; DUSSAP, C. G.; BERTIN, G.; JEMINET, G.; JOUANY, J. P. Alkali extraction of β-D-glucans from Saccharomyces cerevisiae cell wall and study of their adsorptive properties toward zearalenone. Journal of Agricultural and Food Chemistry, v. 52, p. 3666-3673. 2004a.

YIANNIKOURIS, A.; ANDRE, G.; BULEON, A.; JEMINET, G.; CANET, I; FRANCOIS, J.;

BERTIN, G.; JOUANY. J. P. Comprehensive conformational study of key interactions involved in zearalenone complexation with β-d-glucan. Biomacromolecules, v. 5, n. 6, p. 2176- 2185, 2004b.
YIANNIKOURIS, A.; FRANCOIS, J.; POUGHON, L.; DUSSAP, C. G.; JEMINET, G.; BERTIN, G.; JOUANY, J. P. Influence of pH on Complexing of Model b-D-Glucans with Zearalenone. Journal of Food Protection, v. 67, n. 12, p. 2741-2746. 2004.
YIANNIKOURIS, A.; GWÉNAËLLE, A.; POUGHON, L.; FRANCOIS, J.; DUSSAP, C. G.; JEMINET, G.; BERTIN, G.; JOUANY, J. P. Chemical and Conformational Study of the Interactions Involved in Mycotoxin Complexation with β-d-Glucans. Biomacromolecules, v. 7,n. 4, p. 1147-1155. 2006.
YIANNIKOURIS, A.; APAJALAHTI, J.; KETTUNEN, H.; OJANPERÄ, S.; BELL, A. N. W.;KEEGAN, J. D.; MORAN, C. A. Efficient Aflatoxin B1 Sequestration by Yeast CellWall Extract and Hydrated Sodium Calcium Aluminosilicate Evaluated Using a Multimodal In-Vitro and Ex-Vivo Methodology. Toxins, v. 13, p. 24. 2021.
YU, H.: ZHANG, J.; CHEN, Y.; ZHU, J. Zearalenone and Its Masked Forms in Cereals and Cereal-Derived Products: A Review of the Characteristics, Incidence, and Fate in Food Processing. Journal of Fungi, v. 8, n. 9, p. 976. 2022.
ZANGHERI, M.; di NARDO, F.; ANFOSSI, L.; GIOVANNOLI, C.; BAGGIANI, C.; RODA,A.; MIRASOLI, M. A multiplex chemiluminescent biosensor for type B-fumonisins and aflatoxin B1 quantitative detection in maize flour. Analyst, v. 140, p. 358-365. 2015.
ZHANG, Z.; TANG, X.; WANG, D.; ZHANG, Q.; LI, P.; DING, X. Rapid on-site sensing aflatoxin B1 in food and feed via a chromatographic time-resolved fluoroimmunoassay. PLoS ONE, v. 10, p. e0123266. 2015
ZHANG, G.; LI, J.; LV, J.; LIU, L.; LI, C.; LIU, L. Decontamination of aflatoxin M1 in yogurt using Lactobacillus rhamnosus LC-4. Journal of Food Safety, v. 39, p. e12673. 2019.
ZHANG, J.; TANG, X.; CAI, Y.; ZHOU, W.W. Mycotoxin Contamination Status of Cereals in China and Potential Microbial Decontamination Methods. Metabolites, n. 13, p. 551. 2023.
ZHAO, L.; WEI, J.; ZHAO, H.; ZHU, B.; ZHANG, B. Detoxification of carcinogenic compounds by lactic acid bacteria strains. Critical Reviews in Food Science and Nutrition, v. 58, p. 2727-2742. 2018.
ZHAO, L. H.; GUAN, S.; GAO, X.; MA, Q. G.; LEI, Y. P.; BAI, X. M.; JI, C. Preparation, purification and characteristics of an aflatoxin degradation enzyme from Myxococcus fulvus ANSM068. Journal of Applied Microbiology, v. 110, n. 1, p. 147-155. 2011.
ZHENG, J.; GÄNZLE, M. G.; LIN, X. B.; RUAN, L.; SUN, M. Diversity and Dynamics of Bacteriocins from Human Microbiome. Environmental Microbiology, v. 17, p. 2133-2143. 2014.
ZHENG, M. Z.; RICHARD, J. L.; BINDER, J.; A review of rapid methods for the analysis of mycotoxins. Mycopathologia, v. 161, p. 261-273. 2006.
ZINEDINE, A.; SORIANO, J. M.; MOLTÓ, J. C.; MAÑES, J. Review on the toxicity, occurrence, metabolism, detoxification, regulations and intake of zearalenone: An oestrogenic mycotoxin. Food and Chemical Toxicology, v. 45, p. 1-18. 2007.
ZMORA, N.; ZILBERMAN-SCHAPIRA, G.; SUEZ, J.; MOR, U.; DORI-BACHASH, M.; BASHIARDES, S.; KOTLER, E.; ZUR, M.; REGEV-LEHAVI, D.; BRIK, R.B.; FEDERICI, S.; COHEN, Y.; LINEVSKY, R.; ROTHSCHILD, D.; MOOR, A. E.; BEN-MOSHE, S.; HARMELIN, A.; ITZKOVITZ, S.; MAHARSHAK, N.; SHIBOLET, OREN.; SHAPIRO, H.;

PEVSNER-FISHER, M.; SHARON, I.; HALPERN, Z.; SEGAL, E.; ELINAV, E. Personalised gut mucosal colonisation resistance to empiric probiotics is associated with unique host and microbiome features. Cell, v. 174, p. 388-1405. 2018.

ZOGHIA, A.; KHOSRAVI-DARANI, K.; SOHRABVANDIB, S. Surface binding of toxins and heavy metals by probiotics. Mini-Reviews in Medicinal Chemistry, v. 14, p. 84 - 98. 2014.

ZOU, Z.; SUN, J.; HUANG, F.; FENG, Z.; LI, M.; SHI, R.; DING, J.; LI, H. In vitro removal of T-2 toxin by yeasts. Journal of Food Safety, v. 35, p. 544-550. 2015.

Printed by Books on Demand GmbH, Norderstedt / Germany